GRADES 2 - 5

MATH PUZZLERS

25 Reproducible Puzzles, Games, and Activities That Boost the Math Skills and Up the Fun!

by Sonya Kimble-Ellis

SCHOLASTIC
PROFESSIONAL BOOKS

NEW YORK ✦ TORONTO ✦ LONDON ✦ AUCKLAND ✦ SYDNEY

This book is dedicated to Mom, Bernie, Michael
and my loving family.

In memory of my father, John B. Kimble; and grandmother, Daisy Smith.

Special thanks to my editor Virginia Dooley for her support throughout the development of this project; Ingrid Blinken for her keen eye and invaluable contributions to the completion of *Math Puzzlers*; Janice S. Lee for proofreading portions of this manuscript; and Dale Beltzner for testing some of the activities in this text with his students.

Cover design by Jaime Lucero and Vincent Ceci
Cover illustration by Shelley Dieterichs
Interior design by Melinda Belter
Interior illustrations by Melinda Belter and Kate Flanagan

ISBN: 0-590-20943-4

Printed in the U.S.A.

TABLE OF CONTENTS

INTRODUCTION

Welcome to ***Math Puzzlers***! The activities, games and puzzles in this book are designed to help students learn mathematics in a fun yet challenging way. Use them to encourage students to develop computation and problem-solving skills and to aid them in recognizing the need for mathematics in their daily lives.

In addition to mathematical activities, this book provides excellent opportunities to make cross-curricular connections. You may also want to use these activities as a tool to detect your students' strengths and weaknesses in various areas of study.

WHY PUZZLES, GAMES & ACTIVITIES?

Puzzles, games and activities provide perfect opportunities for students to work in groups, interact and communicate with each other and discuss strategies. Puzzles and games require that students use organizational skills and keep records of their scores.

Throughout the text you will find activities that will provide reinforcement for skills your students may have already learned. Each activity has been designed to help you introduce the more difficult skills in a fun and interesting way.

TEACHING TIPS

The activities in this text lend themselves to your needs as a teacher. While some may be used to introduce students to a particular skill, others can be given as extra credit. Though most may require only part of a class session to complete, others may take an entire period. Allow students adequate time to complete each puzzle, game or activity.

You will find that the activities involving easier skills appear in the early part of the book, while the more difficult ones come later. They can be used at any time of year, depending on your schedule or curriculum. The activities are also geared to the NCTM Standards which are described briefly on page 8.

Each activity is proceeded by a page titled "**For The Teacher.**" This page defines the skill or goal of each activity, puzzle, or game and lists the NCTM Standard number or numbers each relates to. There are warm-up activities and math problems, "real-life" connections, and suggested strategies and tips for completing each activity. Each teacher page also includes a reduced student page for quick and easy reference.

In addition to the activity, each student page includes an "**It's Your Turn**" section which challenges students to extend the game or activity in some way.

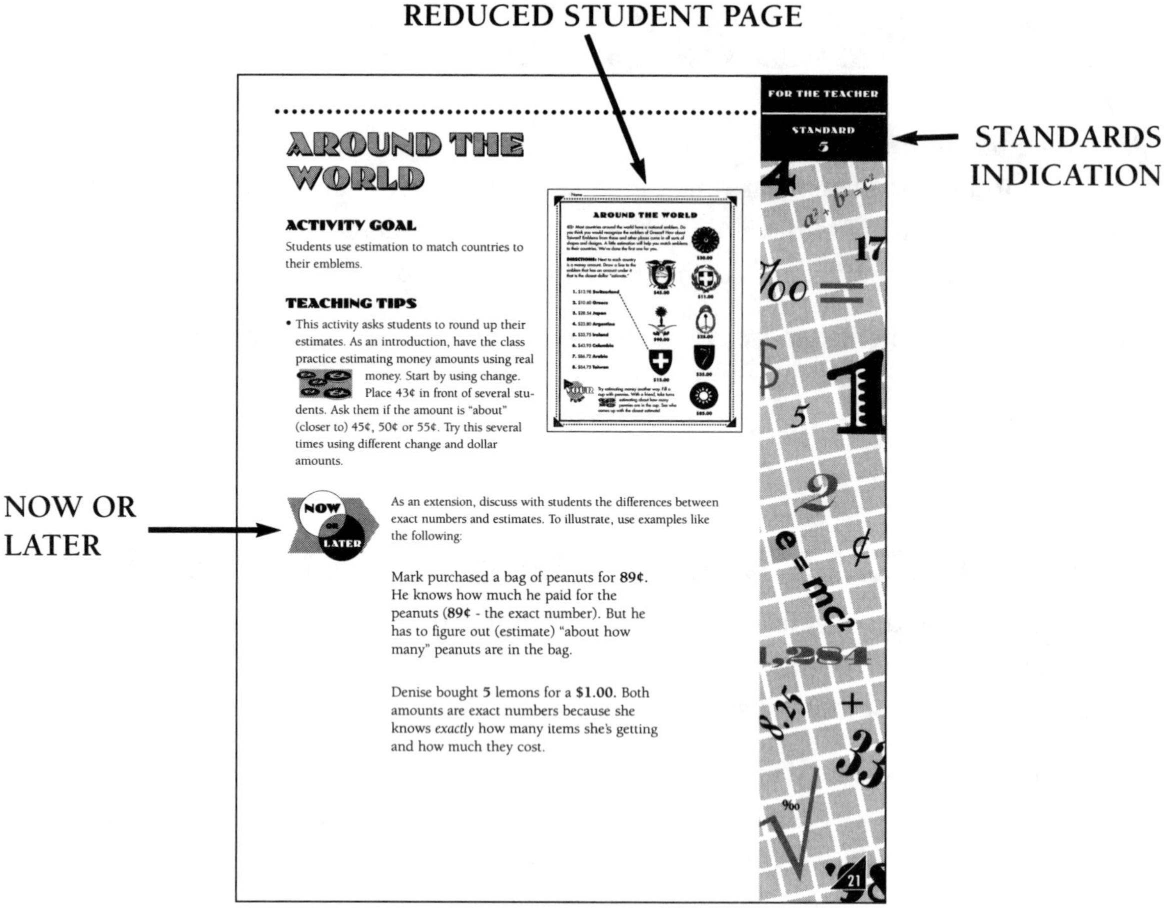

FOR THE TEACHER

STANDARD 5

AROUND THE WORLD

ACTIVITY GOAL

Students use estimation to match countries to their emblems.

TEACHING TIPS

- This activity asks students to round up their estimates. As an introduction, have the class practice estimating money amounts using real money. Start by using change. Place 43¢ in front of several students. Ask them if the amount is "about" (closer to) 45¢, 50¢ or 55¢. Try this several times using different change and dollar amounts.

NOW OR LATER

As an extension, discuss with students the differences between exact numbers and estimates. To illustrate, use examples like the following:

Mark purchased a bag of peanuts for **89¢**. He knows how much he paid for the peanuts (**89¢** - the exact number). But he has to figure out (estimate) "about how many" peanuts are in the bag.

Denise bought **5** lemons for a **$1.00**. Both amounts are exact numbers because she knows *exactly* how many items she's getting and how much they cost.

21

MATERIALS

With the exception of counters, coins and in some instances a calculator, the games require little to no additional materials, so set-up is easy. In most instances, your students will need nothing more than a pencil, as noted by the icon. Where calculators or coins are needed, the following icons will appear:

A FINAL NOTE

It is my hope that your students enjoy completing these puzzles, games and activities—and that they are inspired to work on them not only in the classroom, but at home with their families and friends as well.

NCTM STANDARDS

The National Council of Teachers of Mathematics published 13 Standards that they consider essential to any elementary math curriculum. Each puzzler in this book contains a reference to the NCTM Standard number or numbers it supports. What follows is a brief description of these Standards:

1. **Mathematics as Problem Solving:**
The ability to apply mathematical skills confidently and meaningfully to unfamiliar situations.

2. **Mathematics as Communication:**
Reading, writing, modeling, drawing, and discussing are all important tools in helping children explore, convey, and clarify mathematical concepts and ideas.

3. **Mathematics as Reasoning:**
Giving children opportunities to analyze, draw conclusions, and justify their thinking helps them gain a sense of self-reliance and confidence in their mathematical abilities and helps them see that math makes sense.

4. **Mathematical Connections:**
Helping children see how concepts and ideas in one area of math relate to other areas of math, other subject areas, and their everyday lives fosters an appreciation of the usefulness of mathematics.

5. **Estimation:**
Developing and practicing estimation skills increases children's flexibility in using numbers, equips them with additional skills to apply in reasoning, increases the number and types of situations in which they apply math skills, and helps them realize that mathematics is more than finding exact answers.

6. **Number Sense and Numeration:**
By understanding the meaning of numbers, number relationships, and the effects of operations on numbers, students can make sense of the way numbers are used in the real world.

7. **Concepts of Whole Number Operations:**
Understanding the properties of addition, subtraction, multiplication, and division, knowing when and how to use the operations, and comprehending how the operations relate to one another is the basis for computational skill.

8. **Whole Number Computation:**
Acquiring proficiency with the operations of addition, subtraction, multiplication and division is essential in enabling students to solve problems. Computation methods include paper and pencil and mental math, as well as a calculator or a computer.

9. **Geometry and Spatial Sense:**
The ability to recognize 2- and 3-dimensional shapes, their properties, the relationships among shapes, and to know the effects of changes on shapes helps children understand and describe the physical world around them.

10. **Measurement:**
Students learn and practice measuring to strengthen and formalize their intuitive comparisons of length, width, height, capacity, weight, mass, area, volume, time, temperature, and angle.

11. **Statistics and Probability:**
Collecting, organizing, describing, displaying, and interpreting data, as well as exploring concepts of chance, give students opportunities to investigate and analyze the world around them.

12. **Fractions and Decimals:**
An understanding of fractions and decimals greatly expands students' knowledge of the number system and their appreciation of its usefulness in describing the real world.

13. **Patterns and Relationships:**
The ability to identify patterns and regularities in events, shapes, designs, and sets of numbers, and to describe them mathematically, broadens children's association of mathematics with finding regularities.

STANDARDS
7, 8

PLUS & MINUS PUZZLE

ACTIVITY GOAL

Students add and subtract whole numbers to complete a crossnumber puzzle.

Name

PLUS & MINUS PUZZLE

In this crossnumber puzzle, your mission is to answer these addition and subtraction problems. So you don't get boxed in, we did the first one for you!

243
– 126
117

ACROSS:
A. 243 – 126
C. 96 – 8
E. 105 – 38
F. 18 + 6
G. 65 – 36
H. 43 + 28
I. 234 + 323
K. 53 + 9
L. 84 – 16
N. 134 – 43
O. 80 – 46

DOWN:
A. 455 – 313
B. 41 + 34
C. 624 + 238
D. 5526 + 3264
H. 169 – 92
I. 39 + 17
J. 600 – 71
L. 41 + 22
M. 65 + 19

IT'S YOUR TURN

Using our grid, create your very own crossnumber puzzle. Make up your own addition and subtraction problems. Ask a classmate to complete your puzzle.

9

TEACHING TIPS

- Review place value with the class before doing this activity. Have students look at several groups of numbers. Ask them which numbers are in the ones place, tens place and hundreds place.
- Work on several multi-digit problems with students to ensure that they know "where to start" when adding or subtracting whole numbers. Make sure students understand that they must always start at the right of the problem.
- Also review the rules of "carrying" numbers to complete a multi-digit problem.

DOING THE MATH

➤ADDING WHOLE NUMBERS:

$$\begin{array}{r} {}^{1} \\ 728 \\ +\ 263 \\ \hline 1 \end{array}$$

Start at the right
Add the ones.
8 + 3 = 11
11 = 1 ten and 1 one

$$\begin{array}{r} {}^{1} \\ 728 \\ +\ 263 \\ \hline 91 \end{array}$$

Add the tens
2 + 6 + 1 = 9
There are 9 tens.

$$\begin{array}{r} 728 \\ +\ 263 \\ \hline 991 \end{array}$$

Add the hundreds.
7 + 2 = 9
There are 9 hundreds.

Answer: **991**

➤SUBTRACTING WHOLE NUMBERS:

$$\begin{array}{r} 4 \\ 8\not{5}{}^{1}4 \\ -\ 269 \\ \hline 5 \end{array}$$

Start at the right.
Subtract the ones.
Borrow one from the tens.
14 - 9 = 5
There are 5 ones.

$$\begin{array}{r} 7\ 14 \\ \not{8}\not{5}4 \\ -\ 269 \\ \hline 85 \end{array}$$

Subtract the tens.
Borrow one from the hundreds.
14 - 6 = 8
There are 8 tens.

$$\begin{array}{r} 7 \\ \not{8}54 \\ -\ 269 \\ \hline 585 \end{array}$$

Subtract the hundreds.
7 -2 = 5
There are 5 hundreds.

Answer: **585**

Explain to students that once they master the strategies above, they can also use them when adding or subtracting large numbers with five, six or seven digits.

Name ______________________________

PLUS & MINUS PUZZLE

✏ In this crossnumber puzzle, your mission is to answer these addition and subtraction problems. So you don't get boxed in, we did the first one for you!

$$\begin{array}{r} 243 \\ -\,126 \\ \hline 117 \end{array}$$

ACROSS:

- **A.** 243 – 126
- **C.** 96 – 8
- **E.** 105 – 38
- **F.** 18 + 6
- **G.** 65 – 36
- **H.** 43 + 28
- **I.** 234 + 323
- **K.** 53 + 9
- **L.** 84 – 16
- **N.** 134 – 43
- **O.** 80 – 46

DOWN:

- **A.** 455 – 313
- **B.** 41 + 34
- **C.** 624 + 238
- **D.** 5526 + 3264
- **H.** 169 – 92
- **I.** 39 + 17
- **J.** 600 – 71
- **L.** 41 + 22
- **M.** 65 + 19

A 1	1	B 7	■	C	D
	■		■	E	
F		■	■	G	
■	■	H		■	
I	J		■	■	■
K		■	■	L	M
■	N		■	O	

IT'S YOUR TURN

Using this grid, create your very own crossnumber puzzle. Make up your own addition and subtraction problems. Ask a classmate to complete your puzzle.

STANDARDS 7, 8

ANIMALS MORE OR LESS!

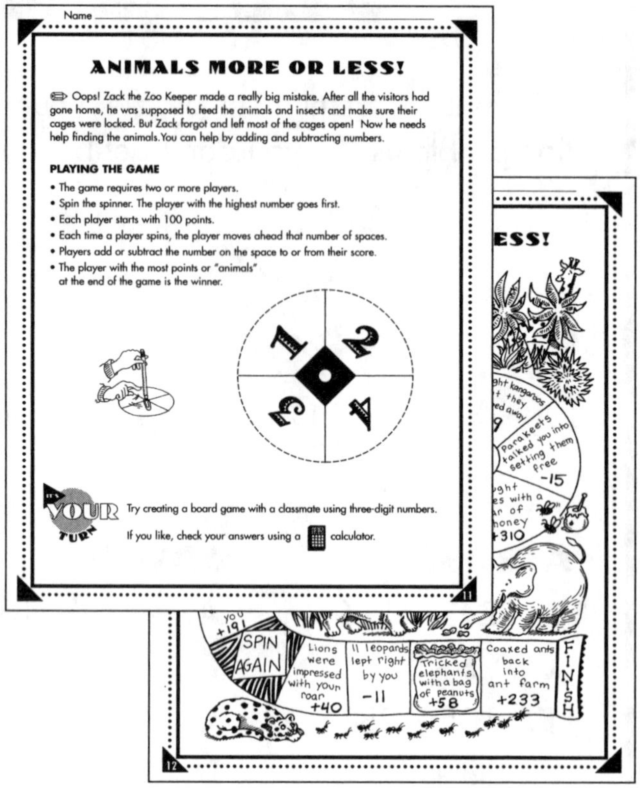
Name

ANIMALS MORE OR LESS!

Oops! Zack the Zoo Keeper made a really big mistake. After all the visitors had gone home, he was supposed to feed the animals and insects and make sure their cages were locked. But Zack forgot and left most of the cages open! Now he needs help finding the animals. You can help by adding and subtracting numbers.

PLAYING THE GAME

- The game requires two or more players.
- Spin the spinner. The player with the highest number goes first.
- Each player starts with 100 points.
- Each time a player spins, the player moves ahead that number of spaces.
- Players add or subtract the number on the space to or from their score.
- The player with the most points or "animals" at the end of the game is the winner.

IT'S YOUR TURN Try creating a board game with a classmate using three-digit numbers.

If you like, check your answers using a calculator.

ACTIVITY GOAL

Students add and subtract numbers to help a zoo keeper find lost animals.

TEACHING TIPS

- Have students use pencil and paper to calculate their score. Once they've played the game, have them play again. This time, challenge students to use "mental math" to determine their score.
- To increase the difficulty of the math, have players start with a score of 500, 1000 or 1500.

NOW OR LATER

Practice is the key to developing whole number addition and subtraction skills. From time to time, have the class complete problems like the ones below. First, ask students to try and find the answer by simply looking at the problem. Then have them work the same problem out on the board. Did they come up with the same answer?

96 + 134 (230)	219 + 26 (245)	1,461 + 323 (1,784)	672 + 68 (740)	2,118 + 241 (2,359)
352 - 29 (323)	126 - 72 (54)	283 - 184 (99)	1,644 - 917 (727)	6,795 - 2,841 (3,954)

Challenge the class to use their addition and subtraction skills to help Zack get the animals back in their cages.

Name ____________________

ANIMALS MORE OR LESS!

Oops! Zack the Zoo Keeper made a really big mistake. After all the visitors had gone home, he was supposed to feed the animals and insects and make sure their cages were locked. But Zack forgot and left most of the cages open! Now he needs help finding the animals. You can help by adding and subtracting numbers.

PLAYING THE GAME

- The game requires two or more players.
- Spin the spinner as shown below. The player with the highest number goes first.
- Each player starts with 100 points.
- Each time a player spins, the player moves ahead that number of spaces.
- Players add or subtract the number on the space to or from their score.
- The player with the most points or "animals" at the end of the game is the winner.

Make a spinner as you see here using the pattern to the right, a paper clip and a pencil.

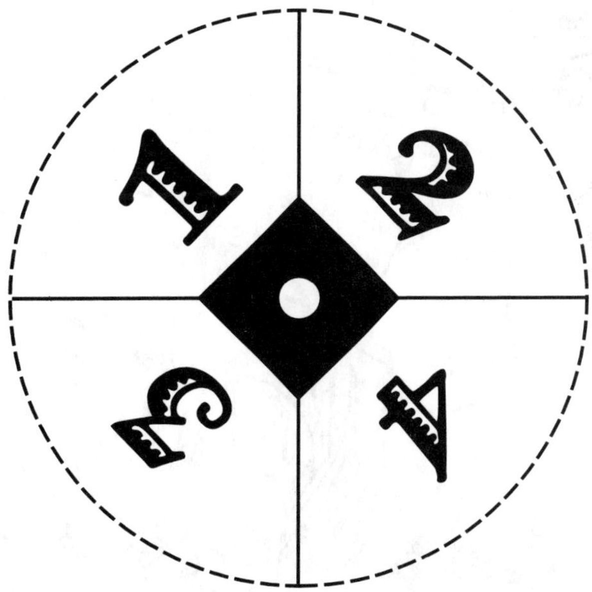

IT'S YOUR TURN Try creating a board game with a classmate using three-digit numbers.

If you like, check your answers using a calculator.

Name ____________________

ANIMALS MORE OR LESS!

STANDARDS
7, 8

TEE-OFF WITH MULTIPLICATION & DIVISION!

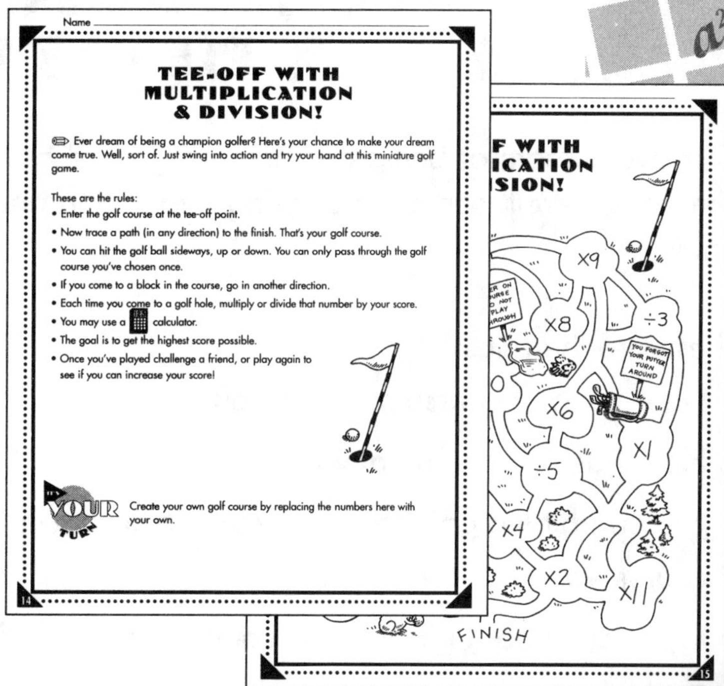
Name

TEE-OFF WITH MULTIPLICATION & DIVISION!

Ever dream of being a champion golfer? Here's your chance to make your dream come true. Well, sort of. Just swing into action and try your hand at this miniature golf game.

These are the rules:
- Enter the golf course at the tee-off point.
- Now trace a path (in any direction) to the finish. That's your golf course.
- You can hit the golf ball sideways, up or down. You can only pass through the golf course you've chosen once.
- If you come to a block in the course, go in another direction.
- Each time you come to a golf hole, multiply or divide that number by your score.
- You may use a calculator.
- The goal is to get the highest score possible.
- Once you've played challenge a friend, or play again to see if you can increase your score!

IT'S YOUR TURN Create your own golf course by replacing the numbers here with your own.

ACTIVITY GOAL

Students use multiplication and division skills to make their way through a miniature golf course.

TEACHING TIPS

- Practicing multiplication tables and division facts is a good introduction or follow-up to this activity. To make practice fun, have students play the following game:

  Write the numbers 1 - 12 on large sheets of paper or cardboard.

1	2	3	4
5	6	7	8
9	10	11	12

Write division and multiplication signs on separate sheets of paper. ÷ X

Ask three students to stand in front of the classroom. Have one student stand in the middle holding the multiplication sign.

Say you'd like to have students multiply numbers by 7. Have one student hold up a card with the number 7 on it. The third student should hold up cards that have other numbers on it. Ask the students in your class to solve each problem.

Have one student record the answers on a large sheet of paper. You can use the answers and the division sign to practice division facts. Once your students have practiced dividing and multiplying, have them play **Tee-Off With Multiplication & Division.**

Name ____________________

TEE-OFF WITH MULTIPLICATION & DIVISION!

Ever dream of being a champion golfer? Here's your chance to make your dream come true. Well, sort of. Just swing into action and try your hand at this miniature golf game.

These are the rules:

- Enter the golf course at the tee-off point.
- Now trace a path (in any direction) to the finish. That's your golf course.
- You can hit the golf ball sideways, up or down. You can only pass through the golf course you've chosen once.
- If you come to a block in the course, go in another direction.
- Each time you come to a golf hole, multiply or divide that number by your score.
- You may use a calculator.
- The goal is to get the highest score possible.
- Once you've played challenge a friend, or play again to see if you can increase your score!

IT'S YOUR TURN

Create your own golf course by replacing the numbers here with your own.

Name ____________________

TEE-OFF WITH MULTIPLICATION & DIVISION!

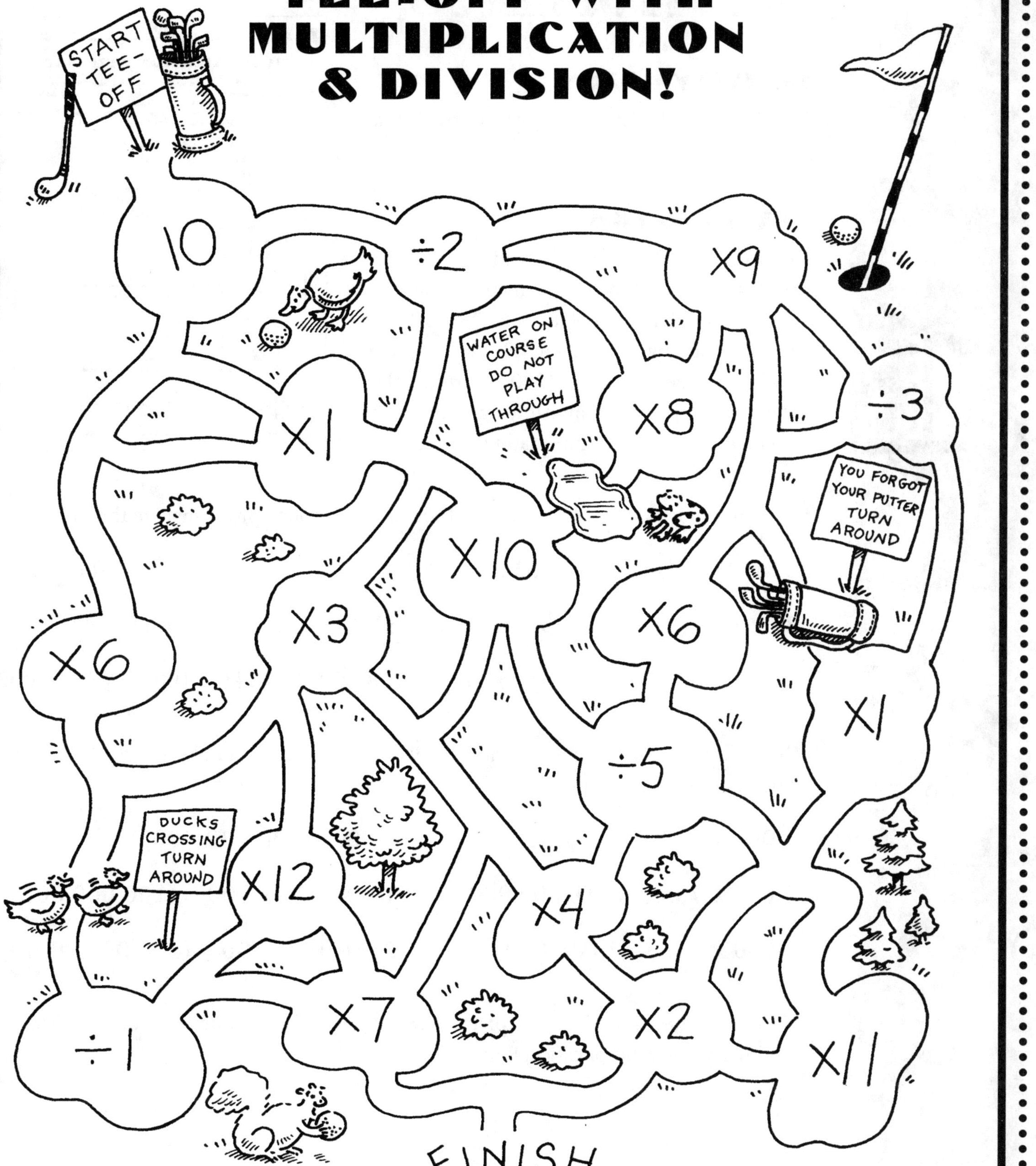

STANDARDS
7, 8

MULTIPLICATION WITH MARTY

ACTIVITY GOAL

Students multiply whole numbers by 10, 100 and 1000.

TEACHING TIPS

- Explain to students that many of the problems in this activity can be solved mentally. Show them how solving a problem like 28 x 100 can be simplified by multiplying 28 x 1 = 28, then adding two zeros (answer: 2800). Explain that the same is true when multiplying numbers by 10 (add one zero) or 1,000 (add three zeros).

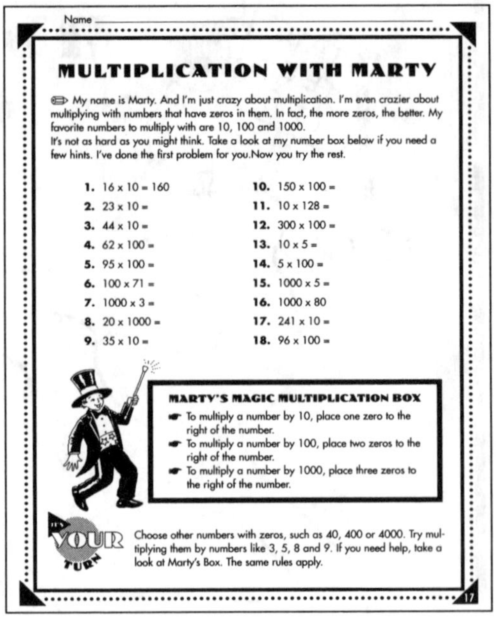

Name

MULTIPLICATION WITH MARTY

My name is Marty. And I'm just crazy about multiplication. I'm even crazier about multiplying with numbers that have zeros in them. In fact, the more zeros, the better. My favorite numbers to multiply with are 10, 100 and 1000.
It's not as hard as you might think. Take a look at my number box below if you need a few hints. I've done the first problem for you.Now you try the rest.

1. 16 x 10 = 160
2. 23 x 10 =
3. 44 x 10 =
4. 62 x 100 =
5. 95 x 100 =
6. 100 x 71 =
7. 1000 x 3 =
8. 20 x 1000 =
9. 35 x 10 =
10. 150 x 100 =
11. 10 x 128 =
12. 300 x 100 =
13. 10 x 5 =
14. 5 x 100 =
15. 1000 x 5 =
16. 1000 x 80
17. 241 x 10 =
18. 96 x 100 =

MARTY'S MAGIC MULTIPLICATION BOX
- To multiply a number by 10, place one zero to the right of the number.
- To multiply a number by 100, place two zeros to the right of the number.
- To multiply a number by 1000, place three zeros to the right of the number.

IT'S YOUR TURN Choose other numbers with zeros, such as 40, 400 or 4000. Try multiplying them by numbers like 3, 5, 8 and 9. If you need help, take a look at Marty's Box. The same rules apply.

17

- To better illustrate this strategy, ask students to complete problems like the following:

34 x 1 = (34)	582 x 1 = (582)	7341 x 1 = (7341)
34 x 10 = (340)	582 x 10 = (5,820)	7341 x 10 = (73,410)
34 x 1000 = (3,400)	582 x 1000 = (582,000)	7341 x 1000 = (7,341,000)

- To challenge the class, guide them in multiplying by larger numbers like 1,000,000. Students can also practice multiplying large numbers such as:

3,000 x 200 = (600,000)	6,000 x 400 = (2,400,000)
3,000 x 7,000 = (21,000,000)	1,000,000 x 100 = (100,000,000)
4,000 x 252 = (1,008,000)	2,000,000 x 50 = (100,000,000)

Once students have had some practice with multiplication, introduce them to Marty and his Magic Multiplication Box.

Name ______________________________

MULTIPLICATION WITH MARTY

My name is Marty. And I'm just crazy about multiplication. I'm even crazier about multiplying with numbers that have zeros in them. In fact, the more zeros, the better. My favorite numbers to multiply with are 10, 100 and 1000.
It's not as hard as you might think. Take a look at my number box below if you need a few hints. I've done the first problem for you. Now you try the rest.

1. 16 x 10 = 160

2. 23 x 10 =

3. 44 x 10 =

4. 62 x 100 =

5. 95 x 100 =

6. 100 x 71 =

7. 1000 x 3 =

8. 20 x 1000 =

9. 35 x 10 =

10. 150 x 100 =

11. 10 x 128 =

12. 300 x 100 =

13. 10 x 5 =

14. 5 x 100 =

15. 1000 x 5 =

16. 1000 x 80

17. 241 x 10 =

18. 96 x 100 =

MARTY'S MAGIC MULTIPLICATION BOX

- To multiply a number by 10, place one zero to the right of the number.
- To multiply a number by 100, place two zeros to the right of the number.
- To multiply a number by 1000, place three zeros to the right of the number.

IT'S YOUR TURN

Choose other numbers with zeros, such as 40, 400 or 4000. Try multiplying them by numbers like 3, 5, 8 and 9. If you need help, take a look at Marty's Box. The same rules apply.

FOR THE TEACHER

STANDARDS
4, 5, 6, 8

MONEY MAGIC PUZZLE

ACTIVITY GOAL

Students round money amounts to the nearest dollar to complete a crossnumber puzzle.

Name

MONEY MAGIC PUZZLE

Whew! Shopping for groceries can be a chore. There are aisles of food to choose from and heavy bags to carry. And of course there's the cost of the groceries in the cart. We can't help with the bags or the choices. But we can give you a few tips for figuring out "about" how much money the groceries will cost. We'll show you one way to round numbers to the nearest dollar when adding money amounts.

DOING THE MATH:

- First, round each price up to the nearest dollar.
- Now add the rounded prices in your head.

$12.98 + $15.69

$12.98 is close to $13

$15.69 is close to $16

Add the two dollar amounts:

$13 + $16 ABOUT $29.00

DIRECTIONS: Circle the correct answer next to each problem on the puzzle page. Write the answer in the crossnumber puzzle.

IT'S YOUR TURN Take a trip to the store with an adult. Try figuring out the total cost of the items before you get to the cash register.

19

						$69	**$63**
						$58	**$42**
					.50	**$521**	**$561**
						$19	**$16**
					.99	**$525**	**$425**
					.99	**$527**	**$566**
						$83	**$75**
M.	$65.75 + $20.90	**$87**	**$97**	**N.**	$39.80 + $13.99	**$54**	**$62**
O.	$9.69 + $32.99	**$40**	**$43**	**O.**	$26.98 + $16.89	**$44**	**$49**
P.	$588.95 + $14.90	**$704**	**$604**	**P.**	$48.95 + $18.99	**$68**	**$66**
Q.	$3.75 + $9.99	**$13**	**$14**				
R.	$428.70 + $50.90	**$480**	**$520**				

20

TEACHING TIPS

- Discuss with students the rules of rounding numbers. Make sure they understand that they should look at the digit to the right of the place they're rounding to. When rounding to the nearest dollar, they should look at the "change" amount (the first digit just to the right of the decimal point).
- Explain to students that the rules of rounding state that if that number is 5 or higher, they should round up. If the number is lower than 5, they should round down.

EXAMPLE: **$11.76.** The first digit to the right of the decimal point is **7**. The number **7** is higher than **5**, so students round **$11.76** up to **$12.00**.

NOW OR LATER

As an introduction or an extension, spend a class session on rounding numbers to the nearest tens, hundreds or thousands place. Give students visual examples.

Round 663 to the nearest hundred.

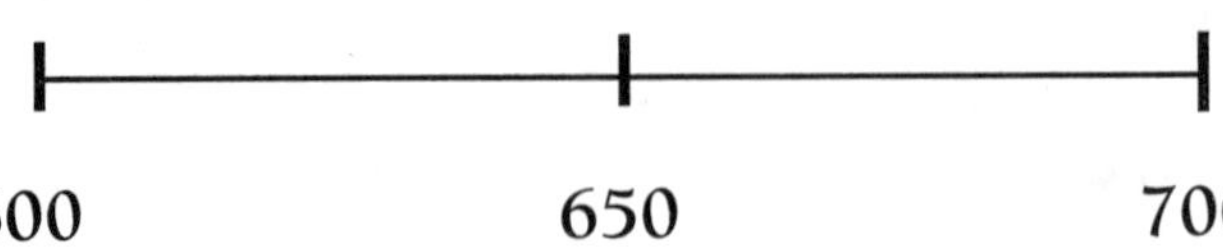

663 is between 600 and 700. But which number is it closer to? 663 is closer to 700. Show students that the 6 (in the tens place) is to the right of the place they're rounding to (the hundreds place). That number (6) is higher than 5, so they should "round up" to 700.

By using this kind of number line as a visual aid, students can see that rounding numbers isn't simply guessing. They must apply a strategy, just as they do when performing other math skills.

Name ____

MONEY MAGIC PUZZLE

Whew! Shopping for groceries can be a chore. There are aisles of food to choose from and heavy bags to carry. And of course there's the cost of the groceries in the cart. We can't help with the bags or the choices. But we can give you a few tips for figuring out "about" how much money the groceries will cost. We'll show you one way to round your total to the nearest dollar.

DOING THE MATH:

- First, round each price up to the nearest dollar.
- Now add the rounded prices in your head.

$12.98 + $15.69

$12.98 is close to $13

$15.69 is close to $16

Add the two dollar amounts:

$13 + $16 = $29.00

So the cost is about $29.00

DIRECTIONS: Circle the correct answer next to each problem on the puzzle page. Write the answer in the crossnumber puzzle.

IT'S YOUR TURN

Take a trip to the store with an adult. Try figuring out the total cost of the items before you get to the cash register.

Name ____________________

MONEY MAGIC PUZZLE

ACROSS:

A.	$16.98 + $18.99	**$36**	**$26**
C.	$24.85 + $29.99	**$65**	**$55**
E.	$21.99 + $8.95	**$31**	**$41**
G.	$218.04 + $67.90	**$286**	**$386**
I.	$53.75 + $40.98	**$105**	**$95**
K.	$7.99 + $19.70	**$28**	**$22**
L.	$99.98 + 99.57	**$300**	**$200**
M.	$65.75 + $20.90	**$87**	**$97**
O.	$9.69 + $32.99	**$40**	**$43**
P.	$588.95 + $14.90	**$704**	**$604**
Q.	$3.75 + $9.99	**$13**	**$14**
R.	$428.70 + $50.90	**$480**	**$520**

DOWN:

B.	$28.59 + $33.95	**$69**	**$63**
C.	$39.25 + $18.70	**$58**	**$42**
D.	$376.35 + $184.50	**$521**	**$561**
F.	$7.28 + $11.69	**$19**	**$16**
H.	$199.80 + $224.99	**$525**	**$425**
J.	$399.95 + $126.99	**$527**	**$566**
M.	$5.85 + $76.95	**$83**	**$75**
N.	$39.80 + $13.99	**$54**	**$62**
O.	$26.98 + $16.89	**$44**	**$49**
P.	$48.95 + $18.99	**$68**	**$66**

AROUND THE WORLD

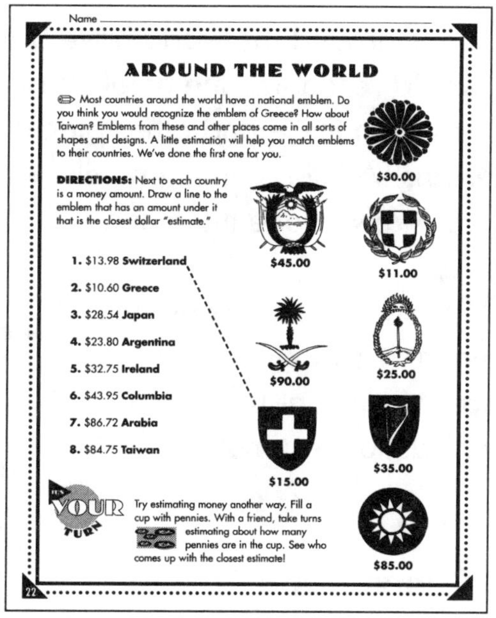

Name ______

AROUND THE WORLD

Most countries around the world have a national emblem. Do you think you would recognize the emblem of Greece? How about Taiwan? Emblems from these and other places come in all sorts of shapes and designs. A little estimation will help you match emblems to their countries. We've done the first one for you.

DIRECTIONS: Next to each country is a money amount. Draw a line to the emblem that has an amount under it that is the closest dollar "estimate."

1. $13.98 **Switzerland**
2. $10.60 **Greece**
3. $28.54 **Japan**
4. $23.80 **Argentina**
5. $32.75 **Ireland**
6. $43.95 **Columbia**
7. $86.72 **Arabia**
8. $84.75 **Taiwan**

$30.00 $45.00 $11.00 $90.00 $25.00 $15.00 $35.00 $85.00

IT'S YOUR TURN Try estimating money another way. Fill a cup with pennies. With a friend, take turns estimating about how many pennies are in the cup. See who comes up with the closest estimate!

22

ACTIVITY GOAL

Students use estimation to match countries to their emblems.

TEACHING TIPS

- This activity asks students to round up their estimates. As an introduction, have the class practice estimating money amounts using real  money. Start by using change. Place 43¢ in front of several students. Ask them if the amount is "about" (closer to) 45¢, 50¢ or 55¢. Try this several times using different change and dollar amounts. Ask students why when handling money, people are more likely to round up, than down.

NOW OR LATER

As an extension, discuss with students the differences between exact numbers and estimates. To illustrate, use examples like the following:

Mark purchased a bag of peanuts for **89¢.** He knows how much he paid for the peanuts (**89¢** - the exact number). But he has to figure out (estimate) "about how many" peanuts are in the bag.

Denise bought **5** lemons for a **$1.00.** Both amounts are exact numbers because she knows *exactly* how many items she's getting and how much they cost.

Name ______________________________

AROUND THE WORLD

Most countries around the world have a national emblem. Do you think you would recognize the emblem of Greece? How about Taiwan? Emblems from these and other places come in all sorts of shapes and designs. A little estimation will help you match emblems to their countries. We've done the first one for you.

DIRECTIONS: Next to each country is a money amount. Draw a line to the emblem that has an amount under it that is the closest dollar "estimate."

1. $13.98 **Switzerland**
2. $10.60 **Greece**
3. $28.54 **Japan**
4. $23.80 **Argentina**
5. $32.75 **Ireland**
6. $43.95 **Colombia**
7. $86.72 **Saudi Arabia**
8. $84.75 **Taiwan**

$30.00

$45.00

$11.00

$90.00

$25.00

$15.00

$35.00

$85.00

IT'S YOUR TURN

Try estimating money another way. Fill a cup with pennies. With a friend, take turns estimating about how many pennies are in the cup. See who comes up with the closest estimate!

TIME FOR PLAY

ACTIVITY GOAL

Students find equivalent measurements to solve a riddle.

Name

TIME FOR PLAY

The dogs in the neighborhood play in the park at the same time every day. Today, some are running around trees and others are playing catch with their owners. But most of them are busy doing something else—chasing another dog! What time were they chasing the dog? Equivalent measurements can help you find the answer.

DIRECTIONS:
- There are two answers next to each question. Circle the letter after the correct answer.
- When you've finished, write each circled letter in the blanks below the riddle. Be sure to write the letters in order.

1. How many weeks are in a year?	34	**L**	52	**T**
2. How many inches are in a foot?	12	**W**	36	**A**
3. How many centimeters are in a meter?	100	**E**	1000	**O**
4. How many nickels are in a dollar?	40	**M**	20	**N**
5. How many days are in a year?	365	**T**	245	**S**
6. How many inches are in a yard?	36	**Y**	24	**B**
7. How many ounces are in a pound?	16	**A**	12	**I**
8. How many hours are in a day?	48	**C**	24	**F**
9. How many years are in a decade?	50	**H**	10	**T**
10. How many cups in a pint?	2	**E**	4	**U**
11. How many quarts are in a gallon?	4	**R**	8	**D**
12. How many feet are in a mile?	5,280	**O**	2,160	**G**
13. How many seconds are in a minute?	30	**J**	60	**N**
14. How many millimeters are in a meter?	1,000	**E**	1500	**P**

What time is it when twenty dogs run after one dog?

______ _____ ___.

IT'S YOUR TURN Come up with an equivalent measurement problem of your own. Exchange problems with a classmate.

24

TEACHING TIPS

- Before beginning this activity, talk to your class about the meaning of the word **equivalent.** Once they have a clear understanding of equivalent measurements, lead them in a discussion about the different ways we measure the same thing. Start by talking to the class about various equivalent measurements for a **month.**
- Have them discuss and answer the following questions:

  How many weeks are in a month? ____________________

  How many days are in the month of March? ____________

  How many hours are there in March?

  ____

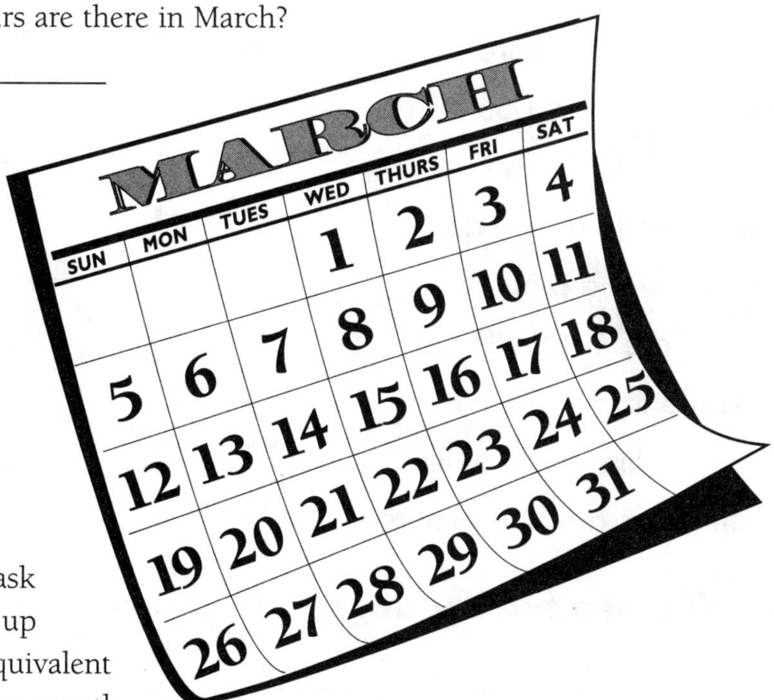

As an extension, ask students to come up with additional equivalent measurements for a month (i.e., minutes in a month, seconds in a month).

Have student test their equivalent measurement skills by solving the riddle of the running dogs.

Name ______________________________

TIME FOR PLAY

The dogs in the neighborhood play in the park at the same time every day. Today, some are running around trees and others are playing catch with their owners. But most of them are busy doing something else—chasing another dog! What time were they chasing the dog? Equivalent measurements can help you find the answer.

DIRECTIONS:

- There are two answers next to each question. Circle the letter after the correct answer.
- When you've finished, write each circled letter in the blanks below the riddle. Be sure to write the letters in order.

1.	How many weeks are in a year?	34	**L**	52	**T**
2.	How many inches are in a foot?	12	**W**	36	**A**
3.	How many centimeters are in a meter?	100	**E**	1000	**O**
4.	How many nickels are in a dollar?	40	**M**	20	**N**
5.	How many days are in a year?	365	**T**	245	**S**
6.	How many inches are in a yard?	36	**Y**	24	**B**
7.	How many ounces are in a pound?	16	**A**	12	**I**
8.	How many hours are in a day?	48	**C**	24	**F**
9.	How many years are in a decade?	50	**H**	10	**T**
10.	How many cups are in a pint?	2	**E**	4	**U**
11.	How many quarts are in a gallon?	4	**R**	8	**D**
12.	How many feet are in a mile?	5,280	**O**	2,160	**G**
13.	How many seconds are in a minute?	30	**J**	60	**N**
14.	How many millimeters are in a meter?	1,000	**E**	1500	**P**

What time is it when twenty dogs run after one dog?

_ _ _ _ _ _   _ _ _ _ _   _ _ _.

IT'S YOUR TURN

Come up with an equivalent measurement problem of your own. Exchange problems with a classmate.

STANDARDS
6, 10

MEASURE BY MEASURE

ACTIVITY GOAL

Students match measuring tools to the things they measure.

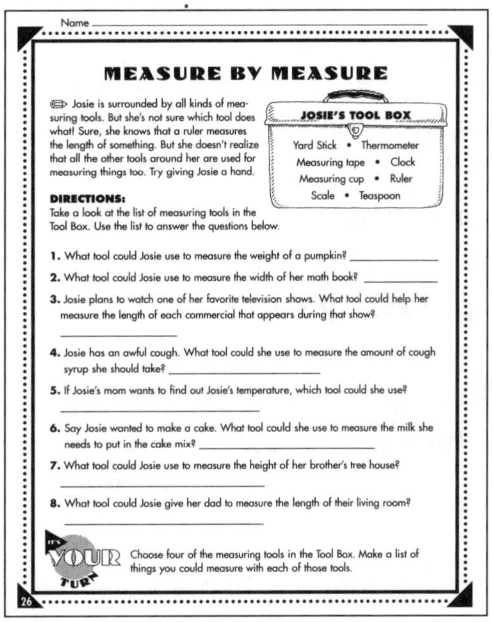
Name

MEASURE BY MEASURE

Josie is surrounded by all kinds of measuring tools. But she's not sure which tool does what! Sure, she knows that a ruler measures the length of something. But she doesn't realize that all the other tools around her are used for measuring things too. Try giving Josie a hand.

JOSIE'S TOOL BOX
Yard Stick • Thermometer
Measuring tape • Clock
Measuring cup • Ruler
Scale • Teaspoon

DIRECTIONS:
Take a look at the list of measuring tools in the Tool Box. Use the list to answer the questions below.

1. What tool could Josie use to measure the weight of a pumpkin? ______
2. What tool could Josie use to measure the width of her math book? ______
3. Josie plans to watch one of her favorite television shows. What tool could help her measure the length of each commercial that appears during that show? ______
4. Josie has an awful cough. What tool could she use to measure the amount of cough syrup she should take? ______
5. If Josie's mom wants to find out Josie's temperature, which tool could she use? ______
6. Say Josie wanted to make a cake. What tool could she use to measure the milk she needs to put in the cake mix? ______
7. What tool could Josie use to measure the height of her brother's tree house? ______
8. What tool could Josie give her dad to measure the length of their living room? ______

IT'S YOUR TURN
Choose four of the measuring tools in the Tool Box. Make a list of things you could measure with each of those tools.

26

TEACHING TIPS

- The word "measurement" encompasses more than measuring the length or width of an object. Explain to students that they can also measure time, temperature and the weight of something. To illustrate this, ask students to participate in a measurement quiz.
- Give them several items to measure and the actual measuring tools (i.e., measuring cup, scale, ruler, clock). Have students guess the measurement of each item first. Then, ask them to actually measure it.
- Students should write **their estimate** and the **actual measurement** on a chart like the one below. They should also calculate the "difference" between their estimate and the actual measurement.

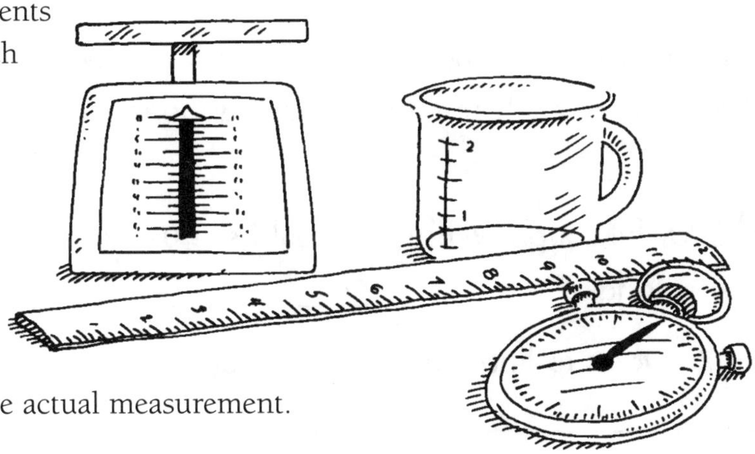

ITEM	ESTIMATE	EXACT MEASUREMENT	DIFFERENCE
Length of our classroom			
Weight of an apple			
Time it takes to count to 50			

Name ____________________

MEASURE BY MEASURE

Josie is surrounded by all kinds of measuring tools. But she's not sure which tool does what! Sure, she knows that a ruler measures the length of something. But she doesn't realize that all the other tools around her are used for measuring things too. Try giving Josie a hand.

JOSIE'S TOOL BOX

Yardstick • Thermometer
Measuring tape • Clock
Measuring cup • Ruler
Scale • Teaspoon

DIRECTIONS:
Take a look at the list of measuring tools in Josie's Tool Box. Use the list to answer the questions below.

1. What tool could Josie use to measure the weight of a pumpkin? ____________________

2. What tool could Josie use to measure the width of her math book? ________________

3. Josie plans to watch one of her favorite television shows. What tool could help her measure the length of each commercial that appears during that show?

____

4. Josie has an awful cough. What tool could she use to measure the amount of cough syrup she should take? ____________________

5. If Josie's mom wants to find out Josie's temperature, which tool could she use?

____

6. Say Josie wanted to make a cake. What tool could she use to measure the milk she needs to put in the cake mix? ____________________

7. What tool could Josie use to measure the height of her brother's tree house?

____

8. What tool could Josie give her dad to measure the length of their living room?

____

IT'S YOUR TURN

Choose four of the measuring tools in Josie's Tool Box. Make a list of things you could measure with each of those tools.

STANDARDS 7, 8, 11

CAT STATS

ACTIVITY GOAL

Students average the scores of costume contest entrants to find the winner.

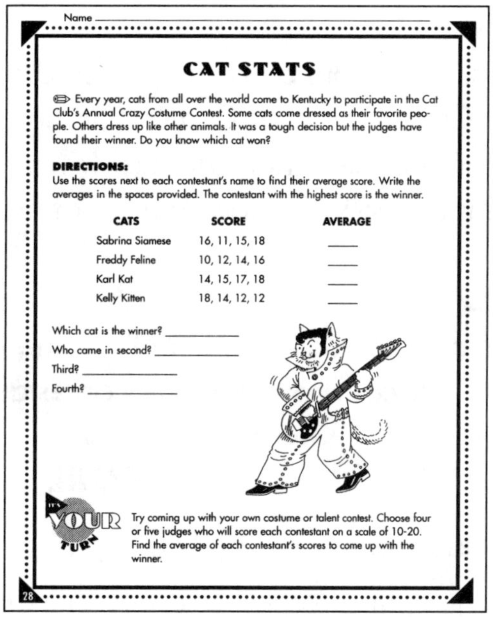

Name ____________________

CAT STATS

Every year, cats from all over the world come to Kentucky to participate in the Cat Club's Annual Crazy Costume Contest. Some cats come dressed as their favorite people. Others dress up like other animals. It was a tough decision but the judges have found their winner. Do you know which cat won?

DIRECTIONS:
Use the scores next to each contestant's name to find their average score. Write the averages in the spaces provided. The contestant with the highest score is the winner.

CATS	SCORE	AVERAGE
Sabrina Siamese	16, 11, 15, 18	____
Freddy Feline	10, 12, 14, 16	____
Karl Kat	14, 15, 17, 18	____
Kelly Kitten	18, 14, 12, 12	____

Which cat is the winner? ____________

Who came in second? ____________

Third? ____________

Fourth? ____________

IT'S YOUR TURN

Try coming up with your own costume or talent contest. Choose four or five judges who will score each contestant on a scale of 10-20. Find the average of each contestant's scores to come up with the winner.

28

TEACHING TIPS

- Averages are used in a variety of ways. Discuss with the class the different contexts in which the word "average" is used. Explain that "average" also refers to mode (the most common number in a set of numbers) or median (the middle number in a set of numbers).
- Before students begin this activity, give them several sets of numbers and ask them to find the averages. Explain that if they were trying to find the average of 16, 8, 5 and 3, they would add all the numbers which gives an answer of 32. Then they would divide 32 by the amount of numbers they added in the first place, 4. So the average of 16, 8, 5, and 3 is equal to $32 \div 4$, or 8.

NOW OR LATER

Averages are used in baseball to calculate batting averages. They are also used to score ice skating competitions. To illustrate a "real-life" connection for the use of averages, have students look at the averages on the back of a baseball card or view an ice skating competition on television.

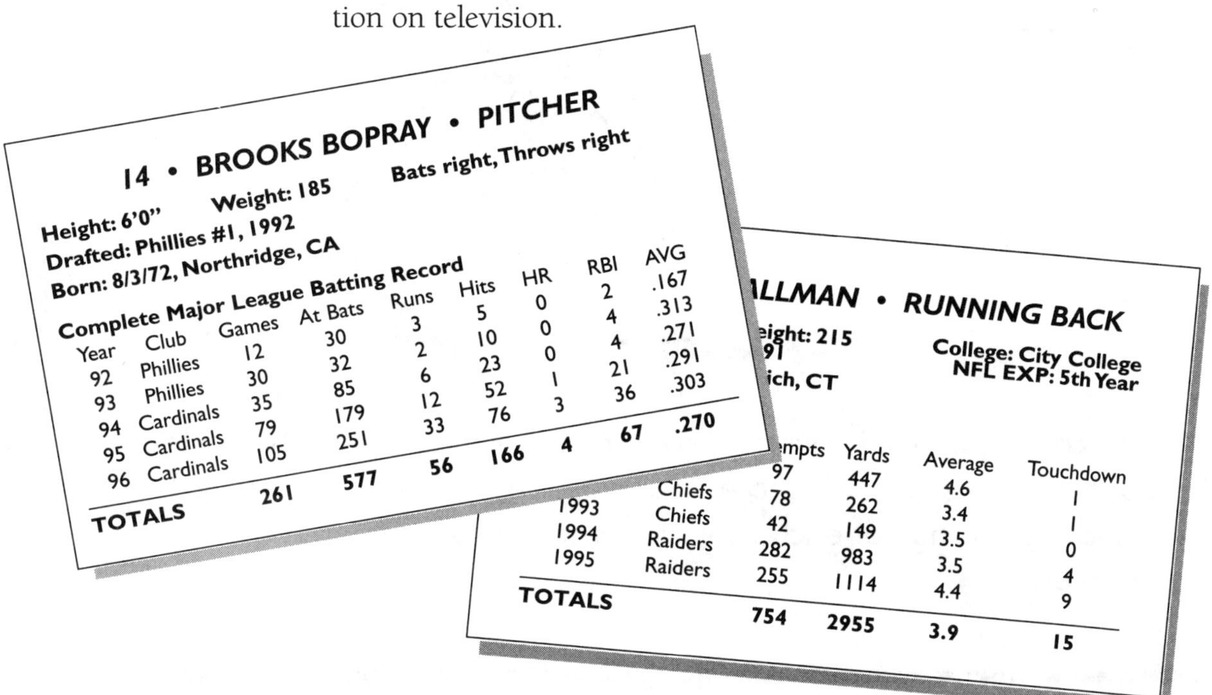

Name ____________________

CAT STATS

Every year, cats from all over the world come to Kentucky to participate in the Cat Club's Annual Crazy Costume Contest. Some cats come dressed as their favorite people. Others dress up like other animals. It was a tough decision, but the judges have found their winner. Do you know which cat won?

DIRECTIONS:
Use the scores next to each contestant's name to find their average score. Write the averages in the spaces provided. The contestant with the highest score is the winner.

CATS	SCORE	AVERAGE
Sabrina Siamese	16, 11, 15, 18	_____
Freddy Feline	10, 12, 14, 16	_____
Karl Kat	14, 15, 17, 18	_____
Kelly Kitten	18, 14, 12, 12	_____

Which cat is the winner? ____________

Who came in second? ____________

Third? ____________

Fourth? ____________

IT'S YOUR TURN

Try coming up with your own costume or talent contest. Choose four or five judges who will score each contestant on a scale of 10-20. Find the average of each contestant's scores to come up with the winner.

STANDARDS
7, 8, 11

PRESIDENTIAL BOARD GAME

ACTIVITY GOAL

Students find the average age of U.S. presidents upon entering the White House.

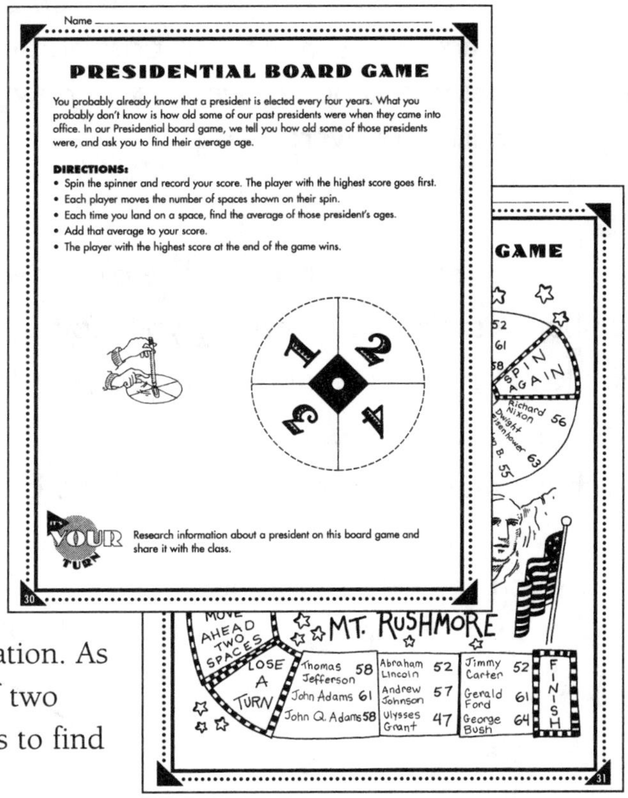
Name

PRESIDENTIAL BOARD GAME

You probably already know that a president is elected every four years. What you probably don't know is how old some of our past presidents were when they came into office. In our Presidential board game, we tell you how old some of those presidents were, and ask you to find their average age.

DIRECTIONS:

- Spin the spinner and record your score. The player with the highest score goes first.
- Each player moves the number of spaces shown on their spin.
- Each time you land on a space, find the average of those president's ages.
- Add that average to your score.
- The player with the highest score at the end of the game wins.

Research information about a president on this board game and share it with the class.

30

TEACHING TIPS

- Before students begin playing the game, explain to them that people often want to find the "average" of a group of numbers that represent something (i.e., the average **height** of a group of people, the average **weight**, an average **distance).**
- Next, demonstrate to the class how averages are used to compare information. As an example, give them the weights of two groups of children. Then ask students to find the average weight for each group.

GROUP A		GROUP B	
Kevin	89	Bryan	84
Ashley	65	Kyle	73
John	72	Sharon	67
Kimberly	76	Michael	81

Which group of children has the higher weight average?
(Group B).

Students can try their hand at averaging other data, such as the average shoe size or average height of several students in the class.

You might also ask them to compare the average age of some of the presidents in the game with the average age of some of the students in the class.

Name ____________________

PRESIDENTIAL BOARD GAME

You probably already know that a president is elected every four years. What you probably don't know is how old some of our past presidents were when they came into office. In our presidential board game, we tell you how old some of those presidents were, and ask you to find their average age.

DIRECTIONS:

- Spin the spinner and record your score. The player with the highest score goes first.
- Move the number of spaces indicated by your spin.
- Each time you land on a space, find the average of those presidents' ages.
- Add that average to your score.
- The player with the highest score at the end of the game wins.

Make a spinner as you see here using the pattern to the right, a paper clip and a pencil.

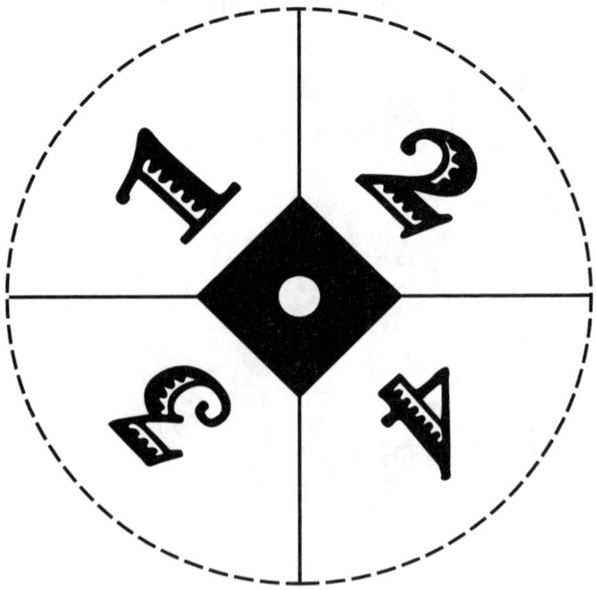

IT'S YOUR TURN

Research information about a president on this game board and share it with the class.

Name ______________________________

PRESIDENTIAL BOARD GAME

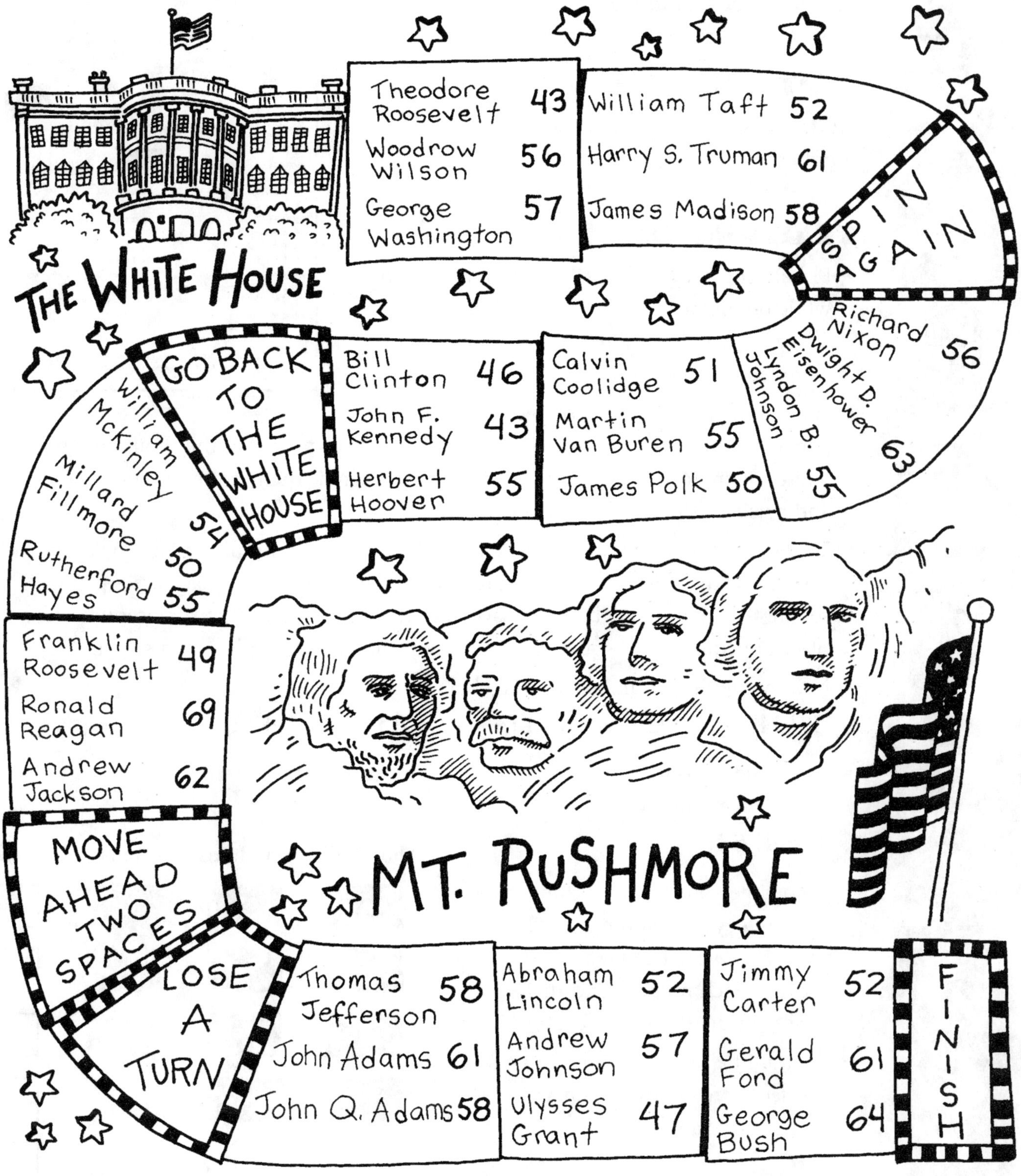

A PLACE FOR EVERY NUMBER PUZZLE

Name ____

A PLACE FOR EVERY NUMBER

Look at the numbers in 243. Each number in the group has its own "place" and meaning. For instance, the 2 in 243 is in the hundreds place. That stands for 2 hundreds or 200. The 4 is in the tens place, meaning 4 tens or 40. And the 3 is in the ones place, meaning 3 ones or 3.

DIRECTIONS:

Use a place value chart to put the numbers in this crossnumber puzzle in their places.

ACROSS

A. 3 hundreds 2 tens 6 ones
C. 8 tens 1 one
E. 6 tens 4 ones
F. 4 tens 7 ones
H. 5 hundreds 2 tens 6 ones
J. 9 tens 3 ones
K. 8 tens 9 ones
M. 5 hundreds 4 tens 2 ones
O. 2 thousands 8 hundreds 3 tens 1 one
Q. 9 tens 8 ones
R. 6 hundreds 6 tens 4 ones

DOWN

A. 3 tens 6 ones
B. 2 thousands 4 hundreds 5 tens 7 ones
D. 1 hundred 4 tens 9 ones
G. 7 tens 3 ones
I. 6 thousands 8 hundreds 5 tens 1one
L. 9 tens 4 ones
N. 2 hundreds 9 tens 6 ones
P. 3 tens 5
R. 8 tens 4 ones

IT'S YOUR TURN Make a list of 10 numbers written out in the same way as the clues above. Ask a classmate to write each of those numbers in their own place value box.

33

ACTIVITY GOAL

Students use place value to complete a cross-number puzzle.

TEACHING TIPS

- Place value is learned easily by some students while others may need more practice. Try the following activity before asking students to complete the puzzle.
- Have students use small colored squares of paper to represent each place. Use a blue square to represent the thousands place, a white square to represent the hundreds place, a red square to represent the tens place, and a yellow square for the ones place.
- To start, each student should have ten squares for each number place. As an introduction, show the class how **ten** squares that represent the ones place are equal to "1" square that represents the tens place; and that **ten** squares that represent the tens place are equal to "1" square that represent the hundreds place and so on.
- Next, write these numbers on the chalk board:

  5,234 8,922 7, 114 3,856 1,482

- Ask students to look at each group of numbers and stack their squares accordingly:

  Ex: 4, 536

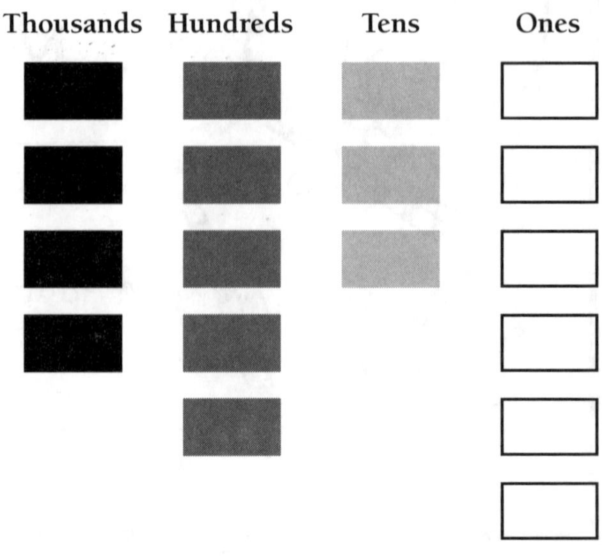

Encourage your students to practice this strategy several times until they make clear visual connections between each number and their "place" in the set of numbers.

After a little practice your students will see there's nothing "square" about place value.

Name ______________________________

A PLACE FOR EVERY NUMBER PUZZLE

Look at the numbers in 243. Each number in the group has its own "place" and meaning. For instance, the 2 in 243 is in the hundreds place. That stands for 2 hundreds or 200. The 4 is in the tens place, meaning 4 tens or 40. And the 3 is in the ones place, meaning 3 ones or 3.

DIRECTIONS:

Use a place value chart to put the numbers in this crossnumber puzzle in their places.

ACROSS

A. 3 hundreds 2 tens 6 ones

C. 8 tens 1 one

E. 6 tens 4 ones

F. 4 tens 7 ones

H. 5 hundreds 2 tens 6 ones

J. 9 tens 3 ones

K. 8 tens 9 ones

M. 5 hundreds 4 tens 2 ones

O. 2 thousands 8 hundreds 3 tens 1 one

Q. 9 tens 8 ones

S. 6 hundreds 6 tens 4 ones

DOWN

A. 3 tens 6 ones

B. 2 thousands 4 hundreds 5 tens 7 ones

D. 1 hundred 4 tens 9 ones

G. 7 tens 3 ones

I. 6 thousands 8 hundreds 5 tens 1 one

L. 9 tens 4 ones

N. 2 hundreds 9 tens 6 ones

P. 3 tens 5

R. 8 tens 4 ones

IT'S YOUR TURN

Make a list of 10 numbers written out in the same way as the clues above. Ask a classmate to write each of those numbers in their own place value box.

STANDARDS
9, 10

GET AN "ANGLE" ON INVENTIONS

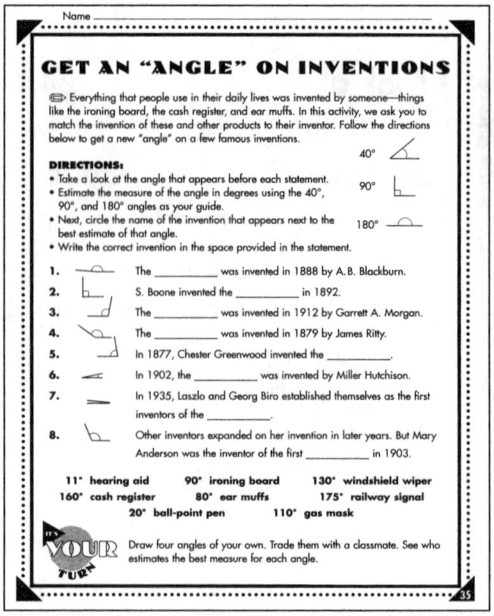

Name ____________

GET AN "ANGLE" ON INVENTIONS

Everything that people use in their daily lives was invented by someone—things like the ironing board, the cash register, and ear muffs. In this activity, we ask you to match the invention of these and other products to their inventor. Follow the directions below to get a new "angle" on a few famous inventions.

40° 90° 180°

DIRECTIONS:
- Take a look at the angle that appears before each statement.
- Estimate the measure of the angle in degrees using the 40°, 90°, and 180° angles as your guide.
- Next, circle the name of the invention that appears next to the best estimate of that angle.
- Write the correct invention in the space provided in the statement.

1. The ________ was invented in 1888 by A. B. Blackburn.
2. S. Boone invented the ________ in 1892.
3. The ________ was invented in 1912 by Garrett A. Morgan.
4. The ________ was invented in 1879 by James Ritty.
5. In 1877, Chester Greenwood invented the ________.
6. In 1902, the ________ was invented by Miller Hutchison.
7. In 1935, Laszlo and Georg Biro established themselves as the first inventors of the ________.
8. Other inventors expanded on her invention in later years. But Mary Anderson was the inventor of the first ________ in 1903.

11° hearing aid · 90° ironing board · 130° windshield wiper · 160° cash register · 80° ear muffs · 175° railway signal · 20° ball-point pen · 110° gas mask

IT'S YOUR TURN: Draw four angles of your own. Trade them with a classmate. See who estimates the best measure for each angle.

35

ACTIVITY GOAL

Students measure the degree of angles to match inventions to their inventors.

TEACHING TIPS

- Explain to students that angles are usually measured in units called degrees. The symbol for degrees is °. Start by showing them that there are 360° in a full circle.

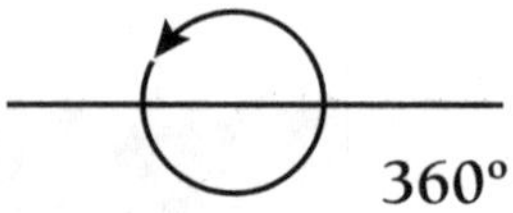

360°

- This will clearly illustrate that the 180° angle, also known as the "straight" angle, is half of a 360° angle.

180°

- Prior to beginning this activity, explain to the class that angles are often measured with a protractor. Try giving students several angles to measure. First, ask them to estimate the degree of those angles. Then, have them measure each angle with a protractor. How close were their estimates?
- Estimates are often made using a "benchmark" or guide. Make sure students understand that the 40°, 90° and 180° angles shown on the activity page are their "benchmarks" for estimating the measure of the angles shown in each statement.

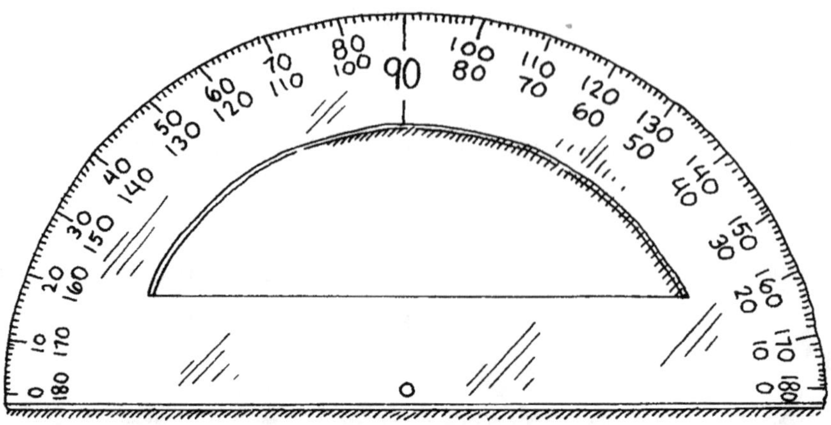

Name ____

GET AN "ANGLE" ON INVENTIONS

Everything that people use in their daily lives was invented by someone—things like the ironing board, the cash register and ear muffs. In this activity, we ask you to match inventions such as these to their inventor. Follow the directions below to get a new "angle" on a few famous inventions.

40° 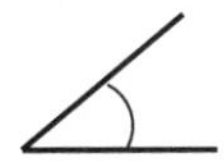

90° 

180° 

DIRECTIONS:

- Take a look at the angle that appears before each statement.
- Estimate the measure of the angle in degrees using the 40°, 90° and 180° angles as your guide.
- Next, circle the name of the invention that appears next to the best estimate of that angle.
- Write the correct invention in the space provided in the statement.

1.  The __________ was invented in 1888 by A.B. Blackburn.

2. 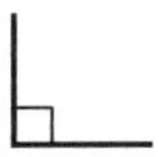 S. Boone invented the __________ in 1892.

3. 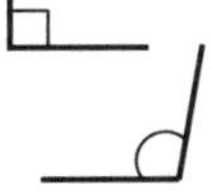 The __________ was invented in 1912 by Garrett A. Morgan.

4.  The __________ was invented in 1879 by James Ritty.

5.  In 1877, Chester Greenwood invented __________.

6.  In 1935, Laszlo and Georg Biro established themselves as the first inventors of the __________.

7.  In 1902, the __________ was invented by Miller Hutchison.

8.  Other inventors expanded on her invention in later years. But Mary Anderson was the inventor of the first __________ in 1903.

11° hearing aid **90° ironing board** **130° windshield wiper**
160° cash register **80° ear muffs** **175° railway signal**
20° ballpoint pen **110° gas mask**

IT'S YOUR TURN

Draw four angles of your own. Trade them with a classmate. See whose estimate is the closest to each angle.

SHAPE UP!

ACTIVITY GOAL

Students answer questions about geometric shapes to solve a riddle.

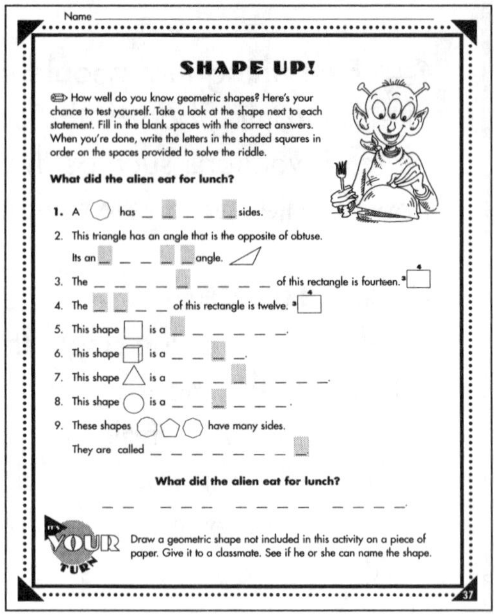
Name

SHAPE UP!

How well do you know geometric shapes? Here's your chance to test yourself. Take a look at the shape next to each statement. Fill in the blank spaces with the correct answers. When you're done, write the letters in the shaded squares in order on the spaces provided to solve the riddle.

What did the alien eat for lunch?

1. A ◯ has _ _ _ _ _ sides.
2. This triangle has an angle that is the opposite of obtuse. Its an _ _ _ _ _ angle.
3. The _ _ _ _ _ _ _ _ _ of this rectangle is fourteen.
4. The _ _ _ _ of this rectangle is twelve.
5. This shape is a _ _ _ _ _ _.
6. This shape is a _ _ _ _.
7. This shape is a _ _ _ _ _ _ _ _.
8. This shape is a _ _ _ _ _ _.
9. These shapes have many sides. They are called _ _ _ _ _ _ _ _.

What did the alien eat for lunch?

_ _ _ _ _ _ _ _ _ _ _ _ _ _

IT'S YOUR TURN Draw a geometric shape not included in this activity on a piece of paper. Give it to a classmate. See if he or she can name the shape.

37

TEACHING TIPS

- Students often think of geometry as complex and difficult. One way to comfortably introduce students to geometry is with familiar shapes. Try making real-life connections for the class by pointing out some geometric shapes they see every day.

A stop sign is the shape of an octagon.

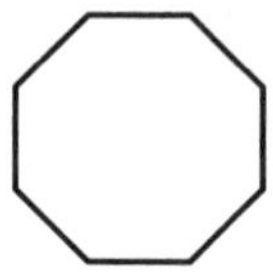

A yield sign is the shape of a triangle.

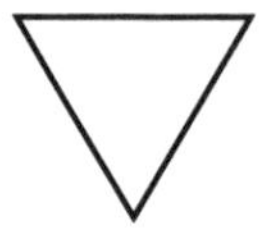

NOW OR LATER Challenge students to find geometric shapes in other objects or items (such as the patterns in their clothing or objects around the house). Ask them to make a list of the items and the geometric shapes they resemble, then share them with the class.

Name ______________________________

SHAPE UP!

How well do you know geometric shapes? Here's your chance to test yourself. Take a look at the shape in each statement. Fill in the blank spaces with the correct answers. When you're done, write the letters in the shaded squares on the spaces provided to solve the riddle.

What did the alien eat for lunch?

1. An  has ___ ___ ___ ___ ___ sides.

2. This triangle has an angle that is the opposite of obtuse.
It's an ___ ___ ___ ___ ___ angle. 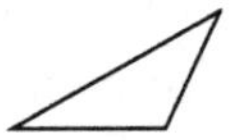

3. The ___ ___ ___ ___ ___ ___ ___ ___ ___ of this rectangle is fourteen. 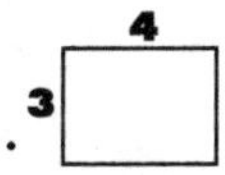

4. The ___ ___ ___ ___ of this rectangle is twelve. 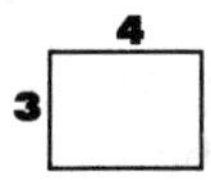

5. This shape  is a ___ ___ ___ ___ ___ ___.

6. This shape 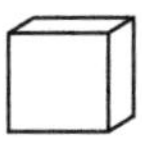 is a ___ ___ ___ ___.

7. This shape  is a ___ ___ ___ ___ ___ ___ ___ ___.

8. This shape 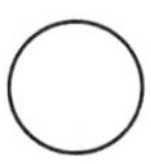 is a ___ ___ ___ ___ ___ ___.

9. These shapes 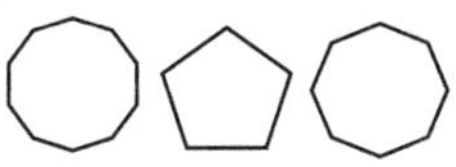 have many sides.
They are called ___ ___ ___ ___ ___ ___ ___ ___.

What did the alien eat for lunch?

___ ___ ___ ___ ___ ___ ___ ___ ___ ___ ___ ___ ___ ___.

IT'S YOUR TURN

Draw a geometric shape not included in this activity on a piece of paper. Give it to a classmate. See if he or she can name the shape.

STANDARDS
6, 7, 8

FACE FACTS

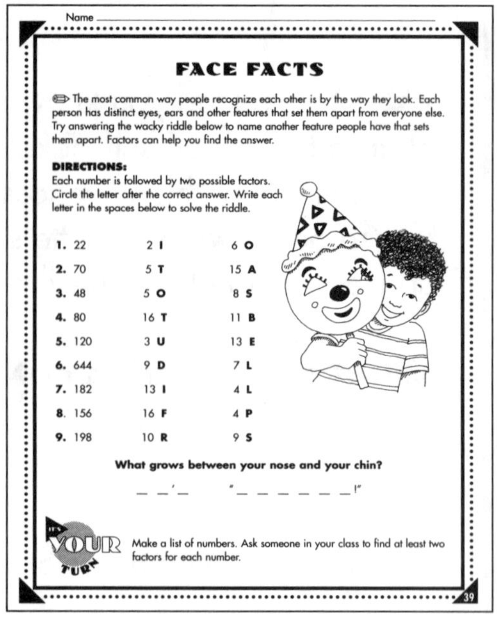

Name ____________

FACE FACTS

The most common way people recognize each other is by the way they look. Each person has distinct eyes, ears and other features that set them apart from everyone else. Try answering the wacky riddle below to name another feature people have that sets them apart. Factors can help you find the answer.

DIRECTIONS:
Each number is followed by two possible factors. Circle the letter after the correct answer. Write each letter in the spaces below to solve the riddle.

1. 22	2 **I**	6 **O**
2. 70	5 **T**	15 **A**
3. 48	5 **O**	8 **S**
4. 80	16 **T**	11 **B**
5. 120	3 **U**	13 **E**
6. 644	9 **D**	7 **L**
7. 182	13 **I**	4 **L**
8. 156	16 **F**	4 **P**
9. 198	10 **R**	9 **S**

What grows between your nose and your chin?

_ _'_ "_ _ _ _ _ _!"

IT'S YOUR TURN Make a list of numbers. Ask someone in your class to find at least two factors for each number.

39

ACTIVITY GOAL

Students find the factors of numbers to solve a riddle.

TEACHING TIPS

- Students can play this fun calculator game as an introduction to finding factors.

1. Give students a number between 2 and 100.
2. Ask them to enter the number on their calculator. Students should then divide it by a (whole) number they think will give them a whole number as their answer.
   If their answer is a whole number, tell them that the number they divided by is a factor of their original number.
3. If students get a whole number, they should continue dividing. They should **STOP** dividing once they get an answer that is not a whole number. Point out that when this happens, the number they divided by is not a factor.
4. Once students understand how to find factors, encourage them to try to find the longest division sequence possible. Tell them they'll earn a point each time they get a whole number as their answer.

   64 ÷ 32 = 2 ÷ 2 = 1 ÷ 1 = 1 **3 points**

   64 ÷ 2 = 32 ÷ 2= 16 ÷ 2 = 8 ÷ 2 = 4 ÷ 2 = 2 ÷ 2 = 1 ÷ 1 = 1 **7 points**

   78 ÷ 2 = 39 ÷ 3 = 13 ÷ 1 = 13 ÷ 13 = 1 ÷ 1 = 1 **5 points**

   50 ÷ 5 = 10 ÷ 2 = 5 ÷ 5 = 1 ÷ 1 = 1 **4 points**

5. The student with the most points wins.

Now that students have had some experience finding factors, challenge them to solve the **Face Facts** riddle.

Name ______________________________

FACE FACTS

The most common way people recognize each other is by the way they look. Each person has distinct eyes, ears and other features that set them apart from everyone else. Try answering the wacky riddle below to name another feature people have that sets them apart. Factors will help you find the answer.

DIRECTIONS:
Each number is followed by two possible factors. Circle the letter after the number that is a factor. Write the letters in order from the first problem to the last to solve the riddle.

1.	22	2 **I**	6 **O**
2.	70	5 **T**	15 **A**
3.	48	5 **O**	8 **S**
4.	80	16 **T**	11 **B**
5.	120	3 **U**	13 **E**
6.	644	9 **D**	7 **L**
7.	182	13 **I**	4 **L**
8.	156	16 **F**	4 **P**
9.	198	10 **R**	9 **S**

What grows between your nose and your chin?

_ _'_ "_ _ _ _ _ _!"

IT'S YOUR TURN

Make a list of numbers. Ask someone in your class to find at least two factors for each number.

STANDARDS
1, 6

SORTING SPORTS

ACTIVITY GOAL

Students complete logic lines to put sports trivia in chronological order.

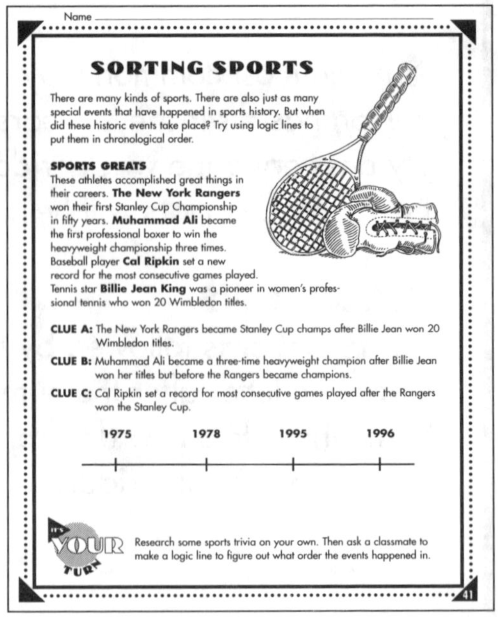

Name

SORTING SPORTS

There are many kinds of sports. There are also just as many special events that have happened in sports history. But when did these historic events take place? Try using logic lines to put them in chronological order.

SPORTS GREATS
These athletes accomplished great things in their careers. **The New York Rangers** won their first Stanley Cup Championship in fifty years. **Muhammad Ali** became the first professional boxer to win the heavyweight championship three times. Baseball player **Cal Ripkin** set a new record for the most consecutive games played. Tennis star **Billie Jean King** was a pioneer in women's professional tennis who won 20 Wimbledon titles.

CLUE A: The New York Rangers became Stanley Cup champs after Billie Jean won 20 Wimbledon titles.

CLUE B: Muhammad Ali became a three-time heavyweight champion after Billie Jean won her titles but before the Rangers became champions.

CLUE C: Cal Ripkin set a record for most consecutive games played after the Rangers won the Stanley Cup.

1975	1978	1995	1996

IT'S YOUR TURN: Research some sports trivia on your own. Then ask a classmate to make a logic line to figure out what order the events happened in.

41

TEACHING TIPS

- Show students how logic lines can be a good way to put dates or events in order. First, select four students in the class. Privately ask each student their birth date. Then draw a large logic line on the board with the birth dates on it.

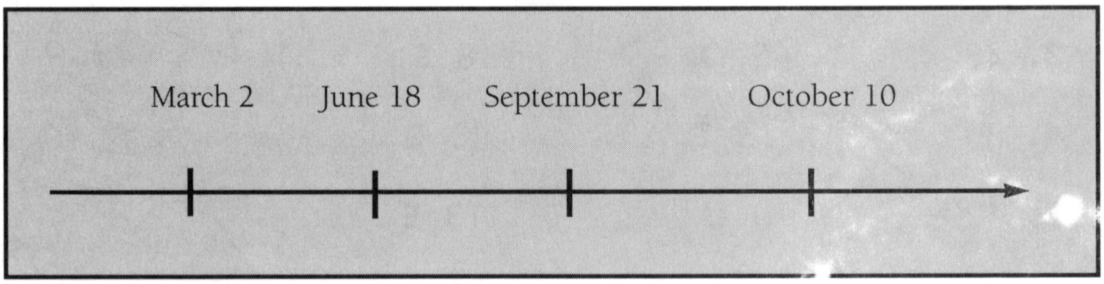

Explain to the class that earlier dates or events are always written toward the left of the line and later ones toward the right.

Next, write several clues on the board such as "Susie's birthday is before Jaime's but after Anna's." Encourage the class to look at them and discuss how they would fill in the logic line.

Now that the class has "ordered" a few birthdays, have them take a swing at some sports trivia.

NOW OR LATER

To tie in other curriculum, students can also make logic lines using important dates in history.

Name ______________________________

SORTING SPORTS

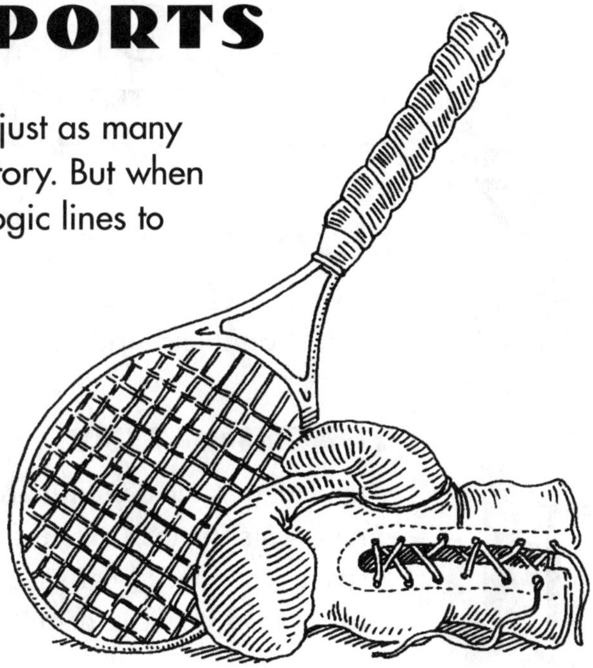

There are many kinds of sports. There are also just as many special events that have happened in sports history. But when did these historic events take place? Try using logic lines to put them in chronological order.

SPORTS GREATS

These athletes accomplished great things in their careers. **The New York Rangers** won their first Stanley Cup Championship in fifty years. **Muhammad Ali** became the first professional boxer to win the heavyweight championship three times. Baseball player **Cal Ripkin Jr.** set a new record for the most consecutive games played. Tennis star **Billie Jean King** was a pioneer in women's professional tennis who won 20 Wimbledon titles.

CLUE A: The New York Rangers became Stanley Cup champs after Billie Jean won 20 Wimbledon titles.

CLUE B: Muhammad Ali became a three-time heavyweight champion after Billie Jean won her titles but before the Rangers became champions.

CLUE C: Cal Ripkin Jr. set a record for most consecutive games played after the Rangers won the Stanley Cup.

1975 **1978** **1995** **1996**

IT'S YOUR TURN

Research some sports trivia on your own. Then ask a classmate to make a logic line to figure out the order in which the events happened.

STANDARD 12

DUCK INTO ACTION WITH FRACTIONS

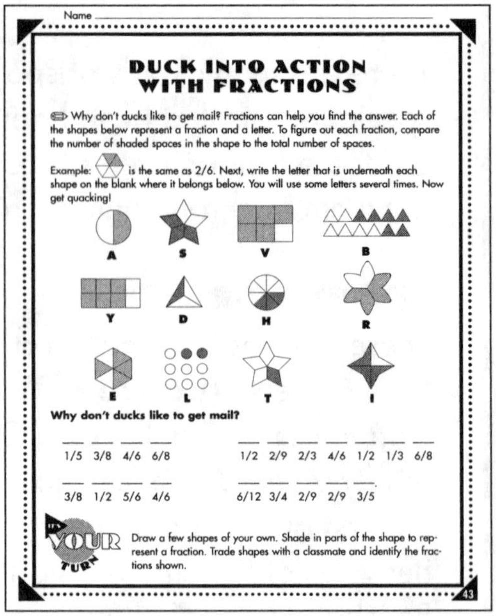
Name

DUCK INTO ACTION WITH FRACTIONS

Why don't ducks like to get mail? Fractions can help you find the answer. Each of the shapes below represent a fraction and a letter. To figure out each fraction, compare the number of shaded spaces in the shape to the total number of spaces.

Example: is the same as 2/6. Next, write the letter that is underneath each shape on the blank where it belongs below. You will use some letters several times. Now get quacking!

A S V B

Y D H R

E L T I

Why don't ducks like to get mail?

1/5 3/8 4/6 6/8 1/2 2/9 2/3 4/6 1/2 1/3 6/8

3/8 1/2 5/6 4/6 6/12 3/4 2/9 2/9 3/5.

IT'S YOUR TURN Draw a few shapes of your own. Shade in parts of the shape to represent a fraction. Trade shapes with a classmate and identify the fractions shown.

43

ACTIVITY GOAL

Students identify the fraction represented in each shape to complete a riddle.

TEACHING TIPS

- Explain that the **denominator** of each fraction represents the total amount of parts in the shape. The shaded parts represent the **numerator**. A food item can help illustrate this strategy.

EXAMPLE:

There are **8** slices of pie shown here **(/8) the denominator**. The shaded area represents how many pieces of the pie you can eat **(1/) the numerator**. The fraction represented in this picture is **1/8.**

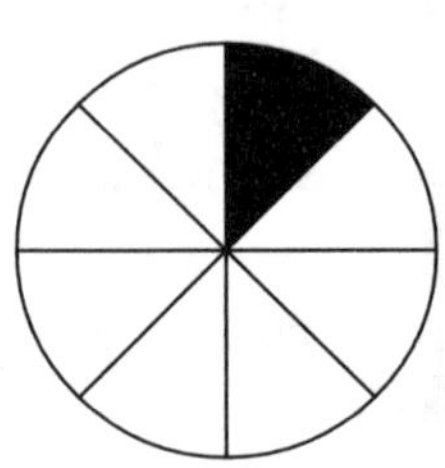

NOW OR LATER

Give students fractioned pictures of other food items. Have them write out the fraction for each item.

Also try a hands-on activity with objects like peanuts or jelly beans. Ask students to fraction those items.

EXAMPLE:

"Take away **1/3** of these **12** items."

Now that their stomachs are probably full, encourage your students to duck into action and name a few fractions to solve the riddle on the following page.

Name ____________________

DUCK INTO ACTION WITH FRACTIONS

Why don't ducks like to get mail? Fractions can help you find the answer. Each of the shapes below represent a fraction and a letter. To figure out each fraction, compare the number of shaded spaces in the shape to the total number of spaces.

Example: 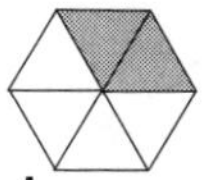 is the same as 2/6. Next, write the letter that is underneath each shape on the corresponding blank below. You will use some letters several times. Now get quacking!

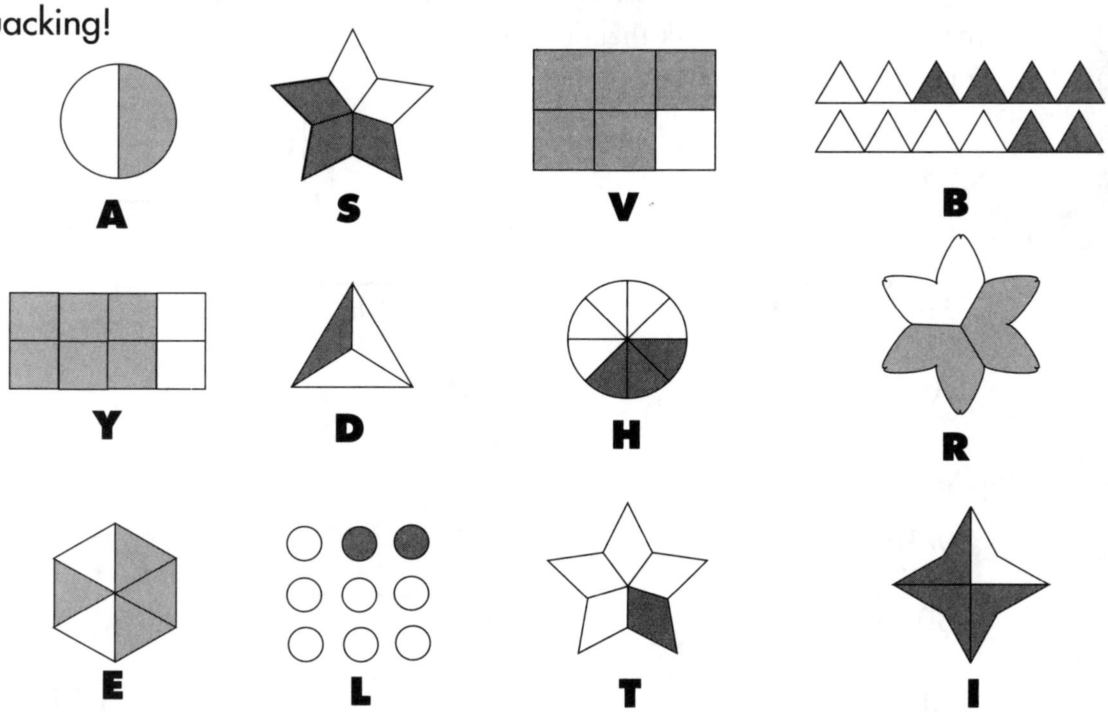

Why don't ducks like to get mail?

___ ___ ___ ___   ___ ___ ___ ___ ___ ___ ___
1/5 3/8 4/6 6/8   1/2 2/9 2/3 4/6 1/2 1/3 6/8

___ ___ ___ ___   ___ ___ ___ ___ ___.
3/8 1/2 5/6 4/6   6/12 3/4 2/9 2/9 3/5

IT'S YOUR TURN

Draw a few shapes of your own. Shade in parts of the shape to represent a fraction. Trade shapes with a classmate and identify the fractions shown.

STANDARDS
7, 12

FRUITY FRACTIONS

ACTIVITY GOAL

Students change fractions into equivalent decimals to solve a riddle.

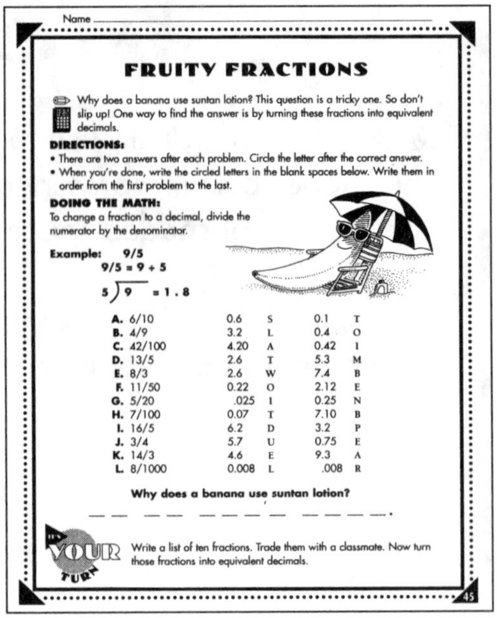

Name ______

FRUITY FRACTIONS

Why does a banana use suntan lotion? This question is a tricky one. So don't slip up! One way to find the answer is by turning these fractions into equivalent decimals.

DIRECTIONS:
- There are two answers after each problem. Circle the letter after the correct answer.
- When you're done, write the circled letters in the blank spaces below. Write them in order from the first problem to the last.

DOING THE MATH:
To change a fraction to a decimal, divide the numerator by the denominator.

Example: 9/5
9/5 = 9 ÷ 5
5) 9 = 1 . 8

A. 6/10	0.6	S	0.1	T
B. 4/9	3.2	L	0.4	O
C. 42/100	4.20	A	0.42	I
D. 13/5	2.6	T	5.3	M
E. 8/3	2.6	W	7.4	B
F. 11/50	0.22	O	2.12	E
G. 5/20	.025	I	0.25	N
H. 7/100	0.07	T	7.10	B
I. 16/5	6.2	D	3.2	P
J. 3/4	5.7	U	0.75	E
K. 14/3	4.6	E	9.3	A
L. 8/1000	0.008	L	.008	R

Why does a banana use suntan lotion?

__ __ __ __ __ __ __ __ __ __ __ __.

IT'S YOUR TURN: Write a list of ten fractions. Trade them with a classmate. Now turn those fractions into equivalent decimals.

45

TEACHING TIPS

- Allow students to use a calculator for this activity. As an introduction, though, ask them to try working out several problems on a sheet of paper. Give the class a list of fractions like the ones below. Then ask them to find each fraction's decimal equivalent.

EXAMPLES:

8/10	6/20
10) 8.000	20) 6.000
(0.8)	(0.3)

PROBLEMS:

9/30	5/40	7/50
30) 9	40) 5	50) 7
(0.3)	(0.125)	(0.14)

Once students have completed the problems, have them check their answers using a calculator.

NOW OR LATER

As an extension, use money amounts to illustrate decimal and fraction equivalents.

1/4 of a dollar = .25 (25¢)
1/2 of a dollar = 0.5 (50¢)

Now that your students have completed the problems above, ask them to solve the mystery of the banana and the suntan lotion.

Name ______________________________

FRUITY FRACTIONS

Why does a banana use suntan lotion? This question is a tricky one. So don't slip up! One way to find the answer is by turning these fractions into equivalent decimals.

DIRECTIONS:

- There are two answers after each problem. Circle the letter after the correct answer.
- When you're done, write the circled letters in the blank spaces below. Write them in order from the first problem to the last.

DOING THE MATH:

To change a fraction to a decimal, divide the numerator by the denominator.

Example: 9/5

9/5 = 9 ÷ 5

$$5\overline{)9} = 1.8$$

A.	6/10	0.6	S	0.1	T
B.	4/9	3.2	L	0.4	O
C.	42/100	4.20	A	0.42	I
D.	13/5	2.6	T	5.3	M
E.	8/3	2.6	W	7.4	B
F.	11/50	0.22	O	2.12	E
G.	5/20	.025	I	0.25	N
H.	7/100	0.07	T	7.10	B
I.	16/5	6.2	D	3.2	P
J.	3/4	5.7	U	0.75	E
K.	14/3	4.6	E	9.3	A
L.	8/1000	0.008	L	.008	R

Why does a banana use suntan lotion?

___ ___   ___ ___   ___ ___ ___ ' ___   ___ ___ ___ ___ .

IT'S YOUR TURN

Write a list of ten fractions. Trade them with a classmate. Now turn those fractions into equivalent decimals.

STANDARDS
6, 7, 8

PRIME ICE CREAM BANDITS

Name ______

PRIME ICE CREAM BANDITS

Irving's Ice Cream Factory is filled with tasty flavors. There's Plum Raisin Supreme, Apple Banana Swirl, Broccoli Chocolate Chunk, and many more. But Irving has a big problem. Some of his workers have been secretly eating pints of their favorite flavors. But who are the culprits? Prime numbers have the answer.

DIRECTIONS:
- Look at the numbers on the shirts of the workers.
- Circle the T-shirts that have prime numbers on them—remember, a prime number is a number that can only be divided by itself and 1 evenly (without a remainder).
- Those workers are the ice cream bandits!

IT'S YOUR TURN Make a list of ten other prime numbers with a classmate.

47

ACTIVITY GOAL

Students find prime numbers to discover who is stealing ice cream from Irving's Ice Cream Factory.

TEACHING TIPS

- Explain to students that a prime number is a number that can only be divided by 1 and itself evenly. Here's a great hands-on way of introducing students to the concept of prime numbers:

Distribute 20 counters to each student and have them create rectangular arrangements for each number from 2 to 20. For example:

2 = 1 x 2 • •
3 = 1 x 3 • • •
4 = 1 x 4 or 2 x 2 • • • • or • •
• •

Ask students to circle the numbers that had only one formation (like 2 and 3 in the above example). Explain that those are prime numbers. You might also ask children if they notice anything else about prime numbers (that all prime numbers with the exception of 2 are odd).

Once students have had the opportunity to identify some prime numbers, challenge them to apply this skill to solving the case of the ice cream bandits.

Name ______________________________

PRIME ICE CREAM BANDITS

Irving's Ice Cream Factory is filled with tasty flavors. There's Plum Raisin Supreme, Apple Banana Swirl, Broccoli Chocolate Chunk, and many more. But Irving has a big problem. Some of his workers have been secretly eating pints of their favorite flavors. But who are the culprits? Prime numbers have the answer.

DIRECTIONS:

- Look at the numbers on the workers' shirts.
- Circle the T-shirts that have prime numbers on them (hint: Try dividing the numbers by those you've already identified as prime.)
- Those workers are the ice cream bandits!

IT'S YOUR TURN

Work with a classmate to make a list of ten more prime numbers.

SIGN IT!

ACTIVITY GOAL

Students write out the numbers shown on street and highway signs as words.

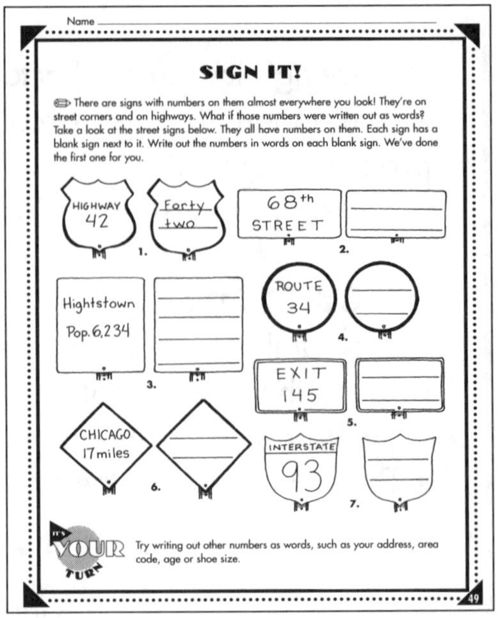
Name

SIGN IT!

There are signs with numbers on them almost everywhere you look! They're on street corners and on highways. What if those numbers were written out as words? Take a look at the street signs below. They all have numbers on them. Each sign has a blank sign next to it. Write out the numbers in words on each blank sign. We've done the first one for you.

IT'S YOUR TURN Try writing out other numbers as words, such as your address, area code, age or shoe size.

49

TEACHING TIPS

- Students should first practice writing out small numbers such as 18 (eighteen) or 23 (twenty-three). Next, have them move on to larger ones like 156, 949 and 2,657.
- Write a list of numbers on the board. Ask students to practice writing those numbers as words on a sheet of paper.

NOW OR LATER

Explain to the class that one "real-life" use for writing out numbers is when people write checks.

Have students take a look at the sample check below. Then give them a blank sample check and a list of money amounts. Ask them to fill in the check, writing the money amounts out as words.

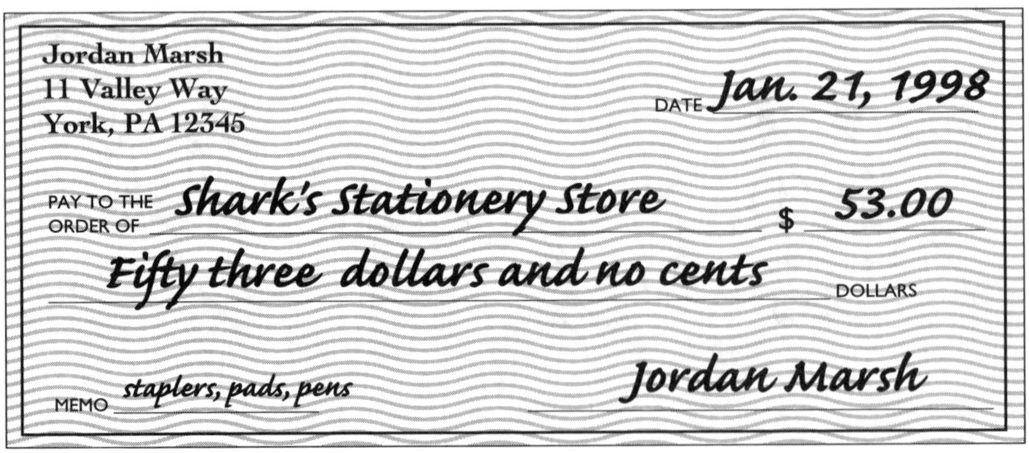

Once students have mastered writing out numbers as words, let them try their hand at changing a few street signs.

Name ______________________________

SIGN IT!

There are signs with numbers on them almost everywhere you look! They're on street corners and on highways. What if those numbers were written out as words? Take a look at the street signs below. They all have numbers on them. Each sign has a blank sign next to it. Write the numbers as words on each blank sign. We've done the first one for you.

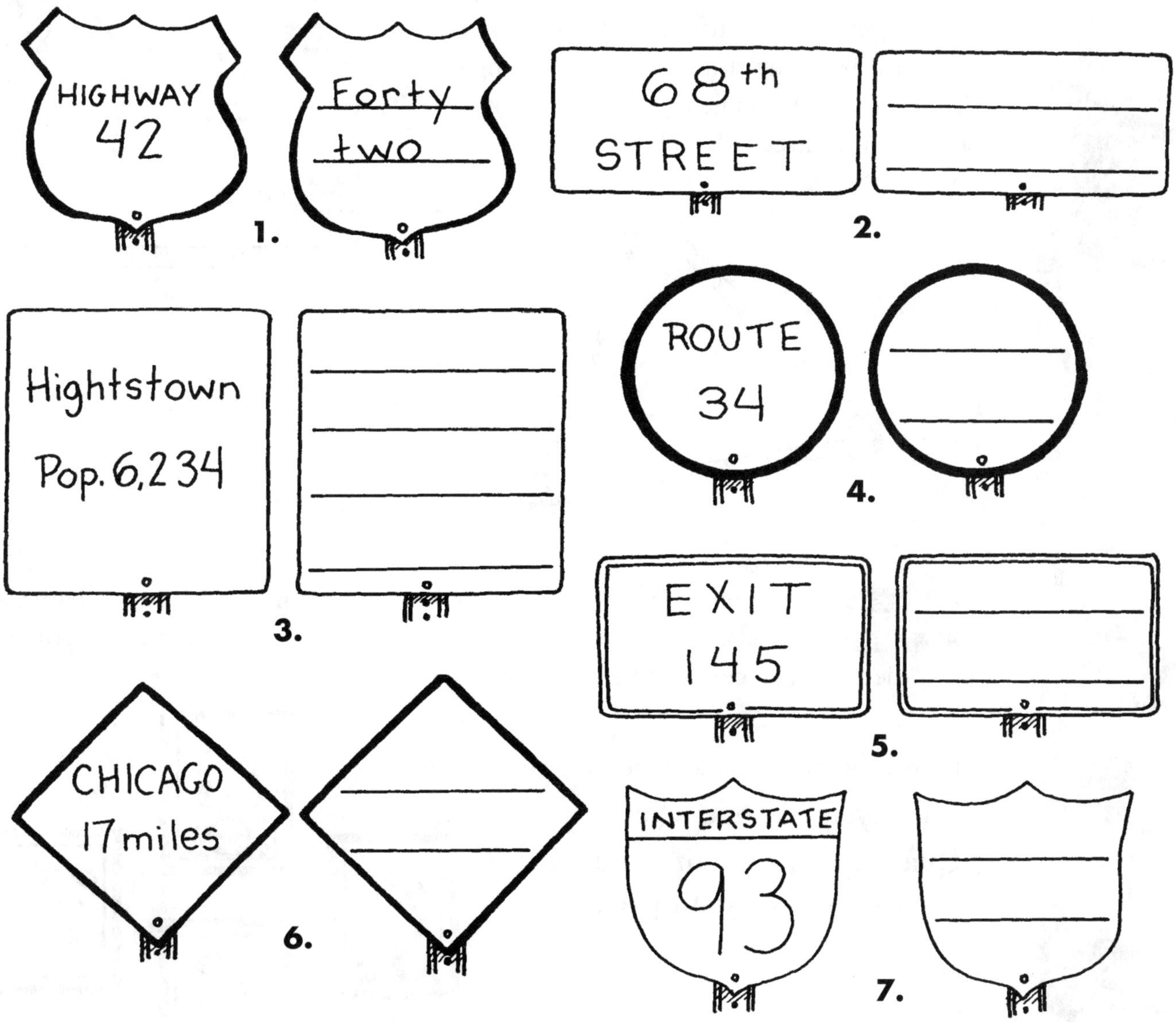

IT'S YOUR TURN

Try writing other numbers as words, such as your address, area code, age or shoe size.

WHAT'S IN A WORD?

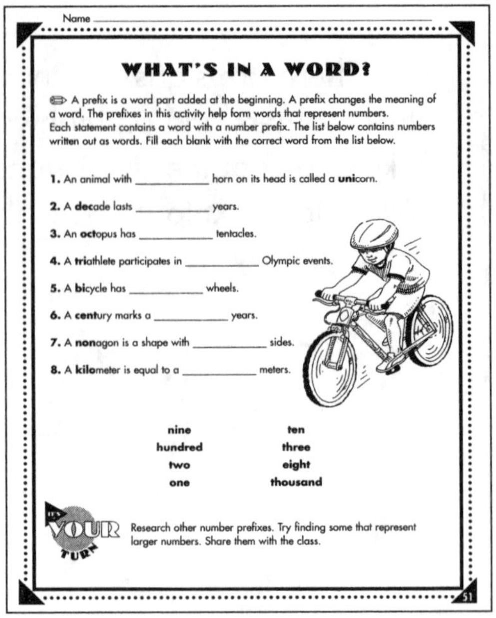
Name ____________

WHAT'S IN A WORD?

A prefix is a word part added at the beginning. A prefix changes the meaning of a word. The prefixes in this activity help form words that represent numbers. Each statement contains a word with a number prefix. The list below contains numbers written out as words. Fill each blank with the correct word from the list below.

1. An animal with ____________ horn on its head is called a **uni**corn.

2. A **dec**ade lasts ____________ years.

3. An **oct**opus has ____________ tentacles.

4. A **tri**athlete participates in ____________ Olympic events.

5. A **bi**cycle has ____________ wheels.

6. A **cent**ury marks a ____________ years.

7. A **non**agon is a shape with ____________ sides.

8. A **kilo**meter is equal to a ____________ meters.

nine	**ten**
hundred	**three**
two	**eight**
one	**thousand**

IT'S YOUR TURN Research other number prefixes. Try finding some that represent larger numbers. Share them with the class.

51

ACTIVITY GOAL

Students complete a statement by matching written out numbers to number prefixes.

TEACHING TIPS

- Use objects to introduce students to number prefixes. Try using straws to make a triangle. This will help illustrate to students that a **tri**angle is a **three**-sided object.

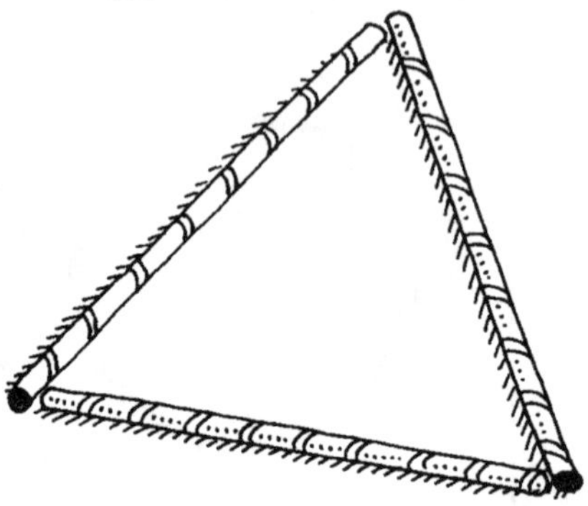

There are 100 cents in a dollar. Use 100 pennies to illustrate that **cent** also means one hundred. As an additional example, explain to students that there are 100 **cent**imeters in a meter.

Here are a few other examples to give students:

A **uni**cycle has **one** wheel.

Quadruplets are **four** children.

An **oct**agon has **eight** sides.

Challenge your students to put these and other prefixes into practice by matching written out numbers to words they've heard many times before.

Name ______________________________

WHAT'S IN A WORD?

A prefix is a word part added at the beginning of a word. A prefix changes the meaning of a word. The prefixes in this activity help form words that represent numbers.

Each statement contains a word with a number prefix. The list below contains numbers written out as words. Fill each blank with the correct word from the list below.

1. An animal with ____________ horn on its head is called a **uni**corn.

2. A **dec**ade lasts ____________ years.

3. An **oct**opus has ____________ tentacles.

4. A **tri**athlete participates in ____________ Olympic events.

5. A **bi**cycle has ____________ wheels.

6. A **cent**ury marks a ____________ years.

7. A **non**agon is a shape with ____________ sides.

8. A **kilo**meter is equal to a ____________ meters.

nine	**ten**
hundred	**three**
two	**eight**
one	**thousand**

IT'S YOUR TURN

Research other number prefixes. Try finding some that represent larger numbers. Share them with the class.

PANSY'S PICTURE PATTERNS

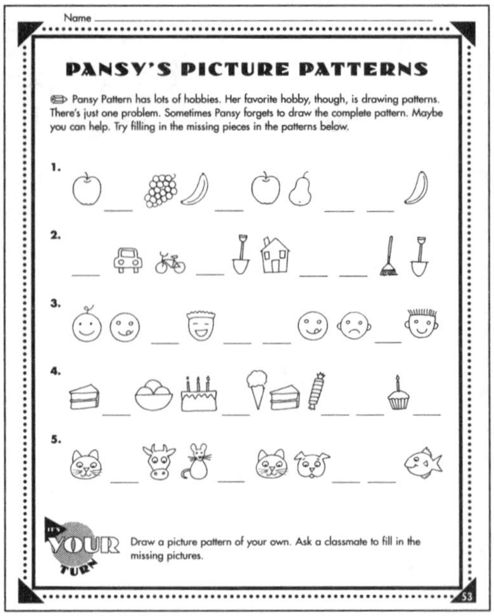
Name

PANSY'S PICTURE PATTERNS

Pansy Pattern has lots of hobbies. Her favorite hobby, though, is drawing patterns. There's just one problem. Sometimes Pansy forgets to draw the complete pattern. Maybe you can help. Try filling in the missing pieces in the patterns below.

1.

2.

3.

4.

5.

IT'S YOUR TURN Draw a picture pattern of your own. Ask a classmate to fill in the missing pictures.

53

ACTIVITY GOAL

Students fill in the blanks to complete a picture pattern.

TEACHING TIPS

- You may want to give students a few hints before they begin working on this activity:
  - ➤ Read each pattern from left to right.
  - ➤ Before trying to identify the pattern, take a look at specific things found in each item in the pattern such as shape, size and number.

You may want to work out the first picture pattern with the class before asking them to complete the activity.

NOW OR LATER

Talk to students about patterns that can be found in nature and architecture such as petal patterns and those found in Islamic and Roman architecture.

As an extension, ask students to visit the local or school library to research patterns and sequences found in ancient architecture.

© Nicholas deVore III / Bruce Colman Inc.

Close-up of tortoise shell

Interior of English cathedral.

© Jim Stamates / Tony Stone Images.

Name ____________________

PANSY'S PICTURE PATTERNS

Pansy Pattern has lots of hobbies. Her favorite hobby, though, is drawing patterns. There's just one problem. Sometimes Pansy forgets to draw the complete pattern. Maybe you can help. Try filling in the missing pieces in the patterns below.

1.

2.

3.

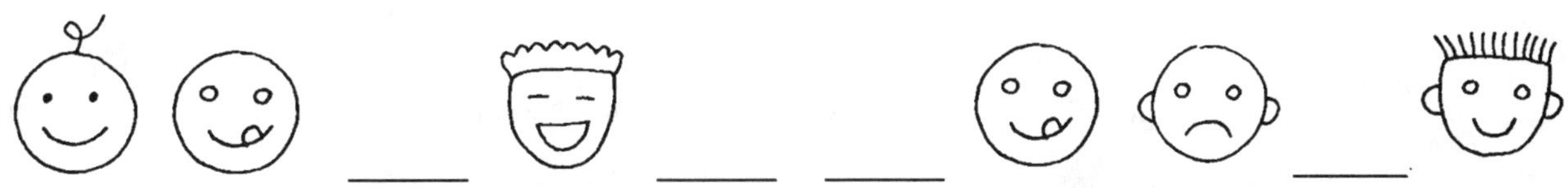

4.

5.

IT'S YOUR TURN

Draw a picture pattern of your own. Ask a classmate to fill in the missing pictures.

THE NEXT NUMBER . . .

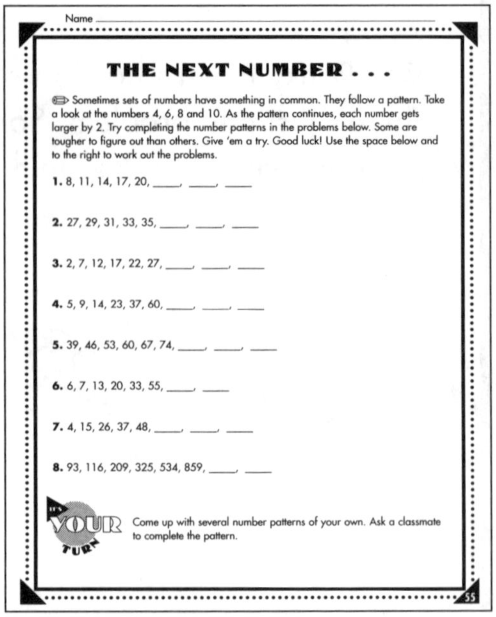
Name ____

THE NEXT NUMBER . . .

Sometimes sets of numbers have something in common. They follow a pattern. Take a look at the numbers 4, 6, 8 and 10. As the pattern continues, each number gets larger by 2. Try completing the number patterns in the problems below. Some are tougher to figure out than others. Give 'em a try. Good luck! Use the space below and to the right to work out the problems.

1. 8, 11, 14, 17, 20, ____, ____, ____

2. 27, 29, 31, 33, 35, ____, ____, ____

3. 2, 7, 12, 17, 22, 27, ____, ____, ____

4. 5, 9, 14, 23, 37, 60, ____, ____, ____

5. 39, 46, 53, 60, 67, 74, ____, ____, ____

6. 6, 7, 13, 20, 33, 55, ____, ____

7. 4, 15, 26, 37, 48, ____, ____, ____

8. 93, 116, 209, 325, 534, 859, ____, ____

IT'S YOUR TURN Come up with several number patterns of your own. Ask a classmate to complete the pattern.

55

ACTIVITY GOAL

Students analyze and complete number patterns.

TEACHING TIPS

- Introduce students to number patterns using the most famous number pattern of all—the **Fibonacci Sequence.**

  First, give students a few numbers in the series:

  1 1 2 3 5 8 13 21 34

Ask students what pattern they see in the numbers. Then ask them to figure out the next four numbers in the pattern (55, 89, 144, 233).

NOW OR LATER

Here's a fun trick students can try with numbers in the series:

- Pick a series of three consecutive numbers: **8, 13, 21**
- Square the number in the middle: **13 x 13 = 169**
- Now multiply the two outside numbers: **8 x 21 = 168**
- Find the difference between the two numbers: **169 - 168 = 1**

Ask students if they think this will this happen with other groups of numbers in the series.

Discuss the Fibonacci Sequence with the students in your class. Explain that the series, named after the man who discovered it, can also be found in things in nature—such as petals on flowers, leaves on stems, and pine cones.

Finding the "next number" isn't always as easy as it seems. Have your students give it a try.

Name ______________________________

THE NEXT NUMBER . . .

Sometimes sets of numbers have something in common. They follow a pattern. Take a look at the numbers 4, 6, 8 and 10. As the pattern continues, each number gets larger by 2. Try completing the number patterns in the problems below. Some are tougher to figure out than others. Give 'em a try. Good luck! Use the space below and to the right to work out the problems.

1. 8, 11, 14, 17, 20, ____, ____, ____

2. 27, 29, 31, 33, 35, ____, ____, ____

3. 2, 7, 12, 17, 22, 27, ____, ____, ____

4. 5, 9, 14, 23, 37, 60, ____, ____, ____

5. 39, 46, 53, 60, 67, 74, ____, ____, ____

6. 6, 7, 13, 20, 33, 55, ____, ____

7. 4, 15, 26, 37, 48, ____, ____, ____

8. 93, 116, 209, 325, 534, 859, ____, ____

IT'S YOUR TURN

Come up with several number patterns of your own. Ask a classmate to complete the pattern.

HOME IMPROVEMENT?

ACTIVITY GOAL

Students find the percents of numbers to complete a story.

TEACHING TIPS

- Here's one way to show students a "real-life" connection for finding percents. Ask them to find the tip amount for several restaurant bills. Explain that tip (gratuity) amounts usually range between 10% and 18%.

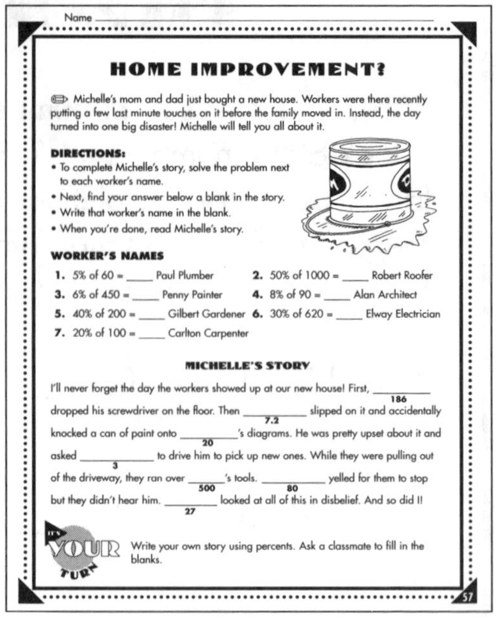
Name ____________

HOME IMPROVEMENT?

Michelle's mom and dad just bought a new house. Workers were there recently putting a few last minute touches on it before the family moved in. Instead, the day turned into one big disaster! Michelle will tell you all about it.

DIRECTIONS:
- To complete Michelle's story, solve the problem next to each worker's name.
- Next, find your answer below a blank in the story.
- Write that worker's name in the blank.
- When you're done, read Michelle's story.

WORKER'S NAMES
1. 5% of 60 = ____ Paul Plumber
2. 50% of 1000 = ____ Robert Roofer
3. 6% of 450 = ____ Penny Painter
4. 8% of 90 = ____ Alan Architect
5. 40% of 200 = ____ Gilbert Gardener
6. 30% of 620 = ____ Elway Electrician
7. 20% of 100 = ____ Carlton Carpenter

MICHELLE'S STORY

I'll never forget the day the workers showed up at our new house! First, ________ (186) dropped his screwdriver on the floor. Then ________ (7.2) slipped on it and accidentally knocked a can of paint onto ________ (20)'s diagrams. He was pretty upset about it and asked ________ (3) to drive him to pick up new ones. While they were pulling out of the driveway, they ran over ______ (500)'s tools. ________ (80) yelled for them to stop but they didn't hear him. ________ (27) looked at all of this in disbelief. And so did I!

IT'S YOUR TURN Write your own story using percents. Ask a classmate to fill in the blanks.

57

EXAMPLES:

1.

BERT'S LUNCH BARN	
1 cheeseburger	$2.43
1 french fry	$1.50
1 soda	.75
	4.68
tax	.28
gratuity 10%	4.96

2.

CHARLOTTES STEAK HOUSE	
2 Sirloin steaks	$16.00
1 bowl of soup	2.35
2 iced teas	3.00
	21.35
tax	1.30
gratuity 15%	22.65

3.

Carla's Cuisine	
shrimp scampi	$19.00
chicken parmegiana	13.00
apple pie	2.95
cheese cake	3.50
2 sodas	4.00
	$42.45
tax	2.50
gratuity 18%	44.95

What tip amounts should the waiter or waitress receive?

1. 50¢ **2. $3.40** **3. $8.09**

Explain to students that they can also find tip amounts using this strategy:

Example: 15% of $80

$$\frac{15}{100} = \frac{x}{80}$$

$$100 \times X = 15 \times 80$$
$$100X = 1200$$
$$X = \frac{1200}{100}$$
$$X = \$12$$

Now challenge students to use this strategy to complete Michelle's story. Once they are done, allow them to check their answers with a calculator.

Name ______________________________

HOME IMPROVEMENT?

Michelle's mom and dad just bought a new house. Workers were there recently putting a few last minute touches on it before the family moved in. But, the day turned into one big disaster! Michelle will tell you all about it.

DIRECTIONS:

- To complete Michelle's story, solve the problem next to each worker's name.
- Next, find your answer below a blank in the story.
- Write that worker's name in the blank.
- When you're done, read Michelle's story.

WORKER'S NAMES

1. 5% of 60 = _____ Paul Plumber
2. 50% of 1000 = _____ Robert Roofer
3. 6% of 450 = _____ Penny Painter
4. 8% of 90 = _____ Alan Architect
5. 40% of 200 = _____ Gilbert Gardener
6. 30% of 620 = _____ Elway Electrician
7. 20% of 100 = _____ Carlton Carpenter

MICHELLE'S STORY

I'll never forget the day the workers showed up at our new house! First, __________ (186) dropped his screwdriver on the floor. Then __________ (7.2) slipped on it and accidentally knocked a can of paint onto __________ (20)'s diagrams. He was pretty upset about it and asked __________ (3) to drive him to pick up new ones. While they were pulling out of the driveway, they ran over _______ (500)'s tools. __________ (80) yelled for them to stop but they didn't hear him. __________ (27) looked at all of this in disbelief. And so did I!

IT'S YOUR TURN

Write your own story using percents. Ask a classmate to fill in the blanks.

STANDARDS
6, 8, 12

WHAT A SALE!

ACTIVITY GOAL

Students use percents to figure out the sale price on items of clothing.

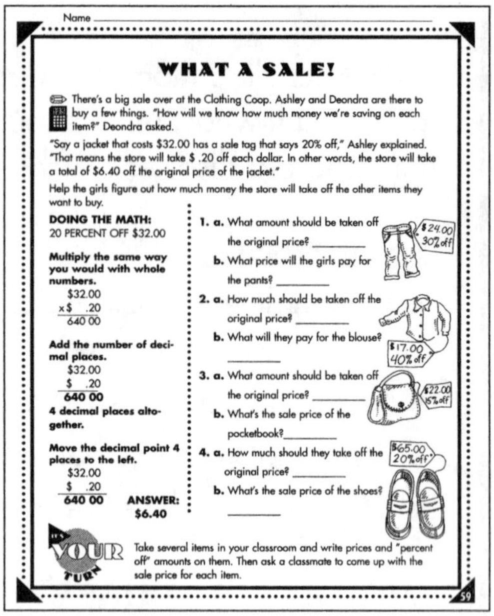
Name

WHAT A SALE!

There's a big sale over at the Clothing Coop. Ashley and Deondra are there to buy a few things. "How will we know how much money we're saving on each item?" Deondra asked.

"Say a jacket that costs $32.00 has a sale tag that says 20% off," Ashley explained. "That means the store will take $ .20 off each dollar. In other words, the store will take a total of $6.40 off the original price of the jacket."

Help the girls figure out how much money the store will take off the other items they want to buy.

DOING THE MATH:
20 PERCENT OFF $32.00

Multiply the same way you would with whole numbers.
$32.00
x$ .20
640 00

Add the number of decimal places.
$32.00
$ .20
640 00
4 decimal places altogether.

Move the decimal point 4 places to the left.
$32.00
$ .20
640 00 **ANSWER: $6.40**

1. a. What amount should be taken off the original price? ________
b. What price will the girls pay for the pants? ________
2. a. How much should be taken off the original price? ________
b. What will they pay for the blouse? ________
3. a. What amount should be taken off the original price? ________
b. What's the sale price of the pocketbook? ________
4. a. How much should they take off the original price? ________
b. What's the sale price of the shoes? ________

IT'S YOUR TURN Take several items in your classroom and write prices and "percent off" amounts on them. Then ask a classmate to come up with the sale price for each item.

59

TEACHING TIPS

- Explain to students that while this activity shows them how to calculate percents on paper, people often find percents using a calculator or by doing "mental math."

  To demonstrate, give students a list of discount problems. Ask a group of students to find the answer using a calculator. Ask another group to solve the same problems using mental math.

EXAMPLE: 20% of $15.00

Show students these steps for discounting a price on a calculator:

What to do	Amounts shown on calculator
Press 15.00	15.00
Press X 10, press %	1.5
Press X 2	3.
Answer: $3.00	

NOW OR LATER

Here's a mental trick students can use to solve the same problem.

To figure out 20% of a number, simply double that number and divide by 10. This strategy is the same as multiplying by .20.

$15.00 becomes $30.00
30 ÷ 10 = 3
Answer: $3.00

Now students can use these strategies to save a little cash at the Clothing Coop sale!

Name ______________________________

WHAT A SALE!

There's a big sale over at the Clothing Coop. Ashley and Deondra are there to buy a few things. "How will we know how much money we're saving on each item?" Deondra asked.

"Say a jacket that costs $32.00 has a sale tag that says 20% off," Ashley explained. "That means the store will take $ .20 off each dollar. In other words, the store will take a total of $6.40 off the original price of the jacket."

Help the girls figure out how much money the store will take off the other items they want to buy.

DOING THE MATH:
20 PERCENT OFF $32.00

Multiply the same way you would with whole numbers.

$32.00
x$ .20
640 00

Add the number of decimal places.

$32.00
$ .20
640 00

4 decimal places altogether.

Move the decimal point 4 places to the left.

$32.00
$ .20
6.4000

ANSWER: $6.40

1. a. What amount should be taken off the original price? __________

b. What price will the girls pay for the pants? __________

2. a. How much should be taken off the original price? __________

b. What will they pay for the blouse? __________

3. a. What amount should be taken off the original price? __________

b. What's the sale price of the pocketbook? __________

4. a. How much should they take off the original price? __________

b. What's the sale price of the shoes? __________

IT'S YOUR TURN

Take several items in your classroom and write prices and "percent off" amounts on them. Then ask a classmate to come up with the sale price for each item.

STANDARDS 4, 6

COUNT LIKE AN EGYPTIAN

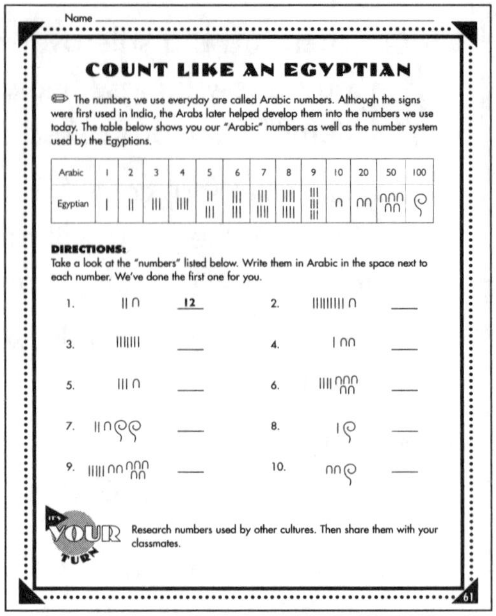
Name ______

COUNT LIKE AN EGYPTIAN

The numbers we use everyday are called Arabic numbers. Although the signs were first used in India, the Arabs later helped develop them into the numbers we use today. The table below shows you our "Arabic" numbers as well as the number system used by the Egyptians.

DIRECTIONS: Take a look at the "numbers" listed below. Write them in Arabic in the space next to each number. We've done the first one for you.

1. ||∩ 12

IT'S YOUR TURN Research numbers used by other cultures. Then share them with your classmates.

61

ACTIVITY GOAL

Students convert Egyptian numerals into Arabic numerals.

TEACHING TIPS

- Give students plenty of time to become acquainted with the Egyptian numerals and their Arabic equivalents as presented in the chart on the following page. Explain that the usual practice in Egypt was to write smaller numbers to the left of the larger ones. The number 15 for example, would be written |||||∩ . You might also want to remind students that Arabic numbers are the numbers we use everyday.

NOW OR LATER

Challenge students to solve addition and subtraction problems using Egyptian numerals.

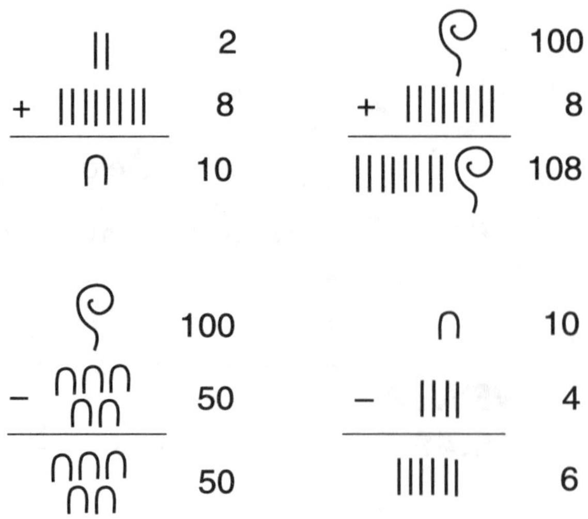

Name ______________________________

COUNT LIKE AN EGYPTIAN

The numbers we use everyday are called Arabic numbers. Although the signs were first used in India, the Arabs later helped develop them into the numbers we use today. The table below shows you our Arabic numbers as well as the number system used by the Egyptians.

<table>
<tr><td>Arabic</td><td>1</td><td>2</td><td>3</td><td>4</td><td>5</td><td>6</td><td>7</td><td>8</td><td>9</td><td>10</td><td>20</td><td>50</td><td>100</td></tr>
<tr><td>Egyptian</td><td>|</td><td>||</td><td>|||</td><td>||||</td><td>||
		</td><td>		
		</td><td>		
			</td><td>			
			</td><td>		
		</td><td>∩</td><td>∩∩</td><td>∩∩∩ ∩∩</td><td>𓍢</td></tr>
</table>

DIRECTIONS:
Take a look at the "numbers" listed below. Write them in Arabic in the space next to each number. We've done the first one for you.

1. ||∩ <u>12</u>

2. |||||||||∩ ____

3. ||||||| ____

4. |∩∩ ____

5. |||∩ ____

6. ||||∩∩∩∩∩ ____

7. ||∩𓍢𓍢 ____

8. |𓍢 ____

9. |||||∩∩∩∩∩∩ ____

10. ∩∩𓍢 ____

IT'S YOUR TURN

Research numbers used by other cultures. Then share them with your classmates.

ANSWERS

PLUS & MINUS PUZZLE

A 1	1	B 7		C 8	D 8
4		5		E 6	7
F 2	4			G 2	9
		H 7	1		0
I 5	J 5	7			
K 6	2			L 6	M 8
	N 9	1		O 3	4

MULTIPLICATION WITH MARTY

1. 160
2. 230
3. 440
4. 6200
5. 9500
6. 7100
7. 3000
8. 20,000
9. 350
10. 150,000
11. 1280
12. 30,000
13. 50
14. 500
15. 5000
16. 80,000
17. 2410
18. 9600

MONEY MAGIC

A 3	B 6				C 5	D 5
	E 3	F 1		G 2	8	6
H 4		I 9	J 5			1
K 2	8		L 2	0	0	
5		M 8	7			N 5
	O 4	3		P 6	0	4
Q 1	4		R 4	8	0	

AROUND THE WORLD

1. $15.00
2. $11.00
3. $30.00
4. $25.00
5. $35.00
6. $45.00
7. $90.00
8. $85.00

TIME FOR PLAY

1. 52
2. 12
3. 100
4. 20
5. 365
6. 36
7. 16
8. 24
9. 10
10. 2
11. 4
12. 5,280
13. 60
14. 1,000

Answer: TWENTY AFTER ONE.

MEASURE BY MEASURE

1. scale
2. ruler
3. clock
4. teaspoon
5. thermometer
6. measuring cup
7. yardstick
8. measuring tape

CAT STATS

Winner - Karl Kat (16)
Second place - Sabrina Siamese (15)
Third place - Kelly Kitten (14)
Fourth place - Freddy Feline (13)

A PLACE FOR EVERY NUMBER

A 3	B 2	6		C 8	D 1	
E 6	4				F 4	G 7
	H 5	2	I 6		J 9	3
	7		K 8	L 9		
			M 5	4	N 2	
O 2	8	P 3	1		Q 9	R 8
		5		S 6	6	4

GET AN "ANGLE" ON INVENTIONS

1. railway signal
2. ironing board
3. gas mask
4. cash register
5. ear muffs
6. ballpoint pen
7. hearing aid
8. windshield wiper

SHAPE UP!

1. eight
2. acute
3. perimeter
4. area
5. square
6. cube
7. triangle
8. circle
9. polygons

Answer: IT ATE MARS BARS.

FACE FACTS

1. 2
2. 5
3. 8
4. 16
5. 3
6. 7
7. 13
8. 4
9. 9

Answer: IT'S "TULIPS!"

SORTING SPORTS

1975	1978	1995	1996
Billie Jean King	Muhammad Ali	NY Rangers	Cal Ripkin Jr.

DUCK INTO ACTION WITH FRACTIONS

T	H	E	Y	A	L	R	E	A	D	Y
1/5	3/8	4/6	6/8	1/2	2/9	2/3	4/6	1/2	1/3	6/8

H	A	V	E	B	I	L	L	S.
3/8	1/2	5/6	4/6	6/12	3/4	2/9	2/9	3/5

FRUITY FRACTIONS

A. 0.6
B. 0.4
C. 0.42
D. 2.6
E. 2.6
F. 0.22
G. 0.25
H. 0.07
I. 3.2
J. 0.75
K. 4.6
L. 0.008

Answer: SO IT WON'T PEEL.

PRIME ICE CREAM BANDITS

The ice cream thieves are wearing shirts with the numbers 11, 19, and 47.

SIGN IT!

2. sixty eighth
3. six thousand two hundred thirty four
4. thirty four
5. one hundred forty five
6. seventeen
7. ninety three

WHAT'S IN A WORD?

1. one
2. ten
3. eight
4. three
5. two
6. hundred
7. nine
8. thousand

PANSY'S PICTURE PATTERNS

1. 

2. 

3. 

4. 

5. 

THE NEXT NUMBER. . .

1. 8, 11, 14, 17, 20, **23**, **26**, **29**
2. 27, 29, 31, 33, 35, **37**, **39**, **41**
3. 2, 7, 12, 17, 22, 27, **32**, **37**, **42**
4. 5, 9, 14, 23, 37, 60, **97**, **157**, **254**
5. 39, 46, 53, 60, 67, 74, **81**, **88**, **95**
6. 6, 7, 13, 20, 33, 55, **88**, **143**
7. 4, 15, 26, 37, 48, **59**, **70**, **81**
8. 93, 116, 209, 325, 534, 859, **1393**, **2252**

HOME IMPROVEMENT?

186 Elway Electrician
7.2 Alan Architect
20 Carlton Carpenter
3 Paul Plumber
500 Robert Roofer
80 Gilbert Gardener
27 Penny Painter

WHAT A SALE!

1a. $7.20
b. $16.80
2a. $6.80
b. $10.20
3a. $3.30
b. $18.70
4a. $13.00
b. $52.00

COUNT LIKE AN EGYPTIAN

1. 12
2. 19
3. 7
4. 21
5. 13
6. 54
7. 212
8. 101
9. 65
10. 120